HTD-Vol. 373
BED-Vol. 53

ADVANCES IN HEAT AND MASS TRANSFER IN BIOTECHNOLOGY – 2002 –

presented at
THE 2002 ASME INTERNATIONAL MECHANICAL ENGINEERING CONGRESS AND EXPOSITION
NOVEMBER 17–22, 2002
NEW ORLEANS, LOUISIANA

sponsored by
THE HEAT TRANSFER DIVISION, ASME
THE BIOENGINEERING DIVISION, ASME

edited by
YILDIZ BAYAZITOGLU
RICE UNIVERSITY

JOHN BISCHOF
UNIVERSITY OF MINNESOTA

LISA XU
PURDUE UNIVERSITY

ELAINE SCOTT
VIRGINIA POLYTECHNIC INSTITUTE AND STATE UNIVERSITY

THE AMERICAN SOCIETY OF MECHANICAL ENGINEERS
Three Park Avenue / New York, N.Y. 10016

ISBN No. 0-7918-3639-8

FOREWORD

The main meeting of the K-17 Committee on Heat and Mass Transfer in Biotechnology occurs annually at the ASME International Mechanical Engineering Congress and Exposition (IMECE). Heat and mass transfer, indeed energy transfer, plays an important role in all biological systems. Applications vary from a variety of thermal therapies, including cryosurgery to biostabilization and cryopreservation as well as fundamental work in cryo and burn injury. The K-17 Committee in ASME actively pursues research at the forefront of this important area and, with the help of both HTD and BED sponsorship, we have eight sessions this year. As is often the case in biomedical engineering, the topics range from fundamental to applied, as the session titles and brief descriptions below will attest:

- **Measurement of Thermal and Mechanical Properties of Biomaterial**. Properties at the extremes of temperature and concentration are needed for accurate predictive models in a range of biomedical applications.
- **Metrics of Thermal Injury**. Accumulation of thermal injury at various micro and macro levels within the system often determine the success of various applications.
- **Heat Transfer in Medicine and Biology: Analysis and Applications**. Thermal therapies (cryosurgery and high temperature therapies) and fundamentals of heat transfer in blood-perfused tissues are important applications highlighted in these sessions.
- **Instrumentation and Measurement in Bio-Heat and Mass Transfer**. Fluorescent and skin flap tools are presented to better understand various transport applications.
- **Modeling of Microscale Bioheat and Mass Transfer**. This session is dedicated to microscale modeling of transport issues, particularly during freeze/thaw applications.
- **Preservation of Tissue Engineered Equivalents**.The successful storage, banking and shipping of tissue-engineered products is dependent on preservation techniques.
- **Micro and Functional Imaging**.This session is dedicated to microscale imaging in both intravital and in vitro systems.

We would like to thank all of the authors, reviewers, session chairs and co-chairs without whose hard work and investment this publication would not have been possible. We would also like to especially thank both the HTD and BED representatives who have continued to support K-17 this year, as in the past, as an interdisciplinary committee within ASME. We hope this collection serves as a useful reference to current work in this exciting and important biomedical area.

Yildiz Bayazitoglu
Heat Transfer Division
Technical Program Representative, IMECE2002
Rice University

John Bischof
K-17 Committee Chair
University of Minnesota

Lisa Xu
K-17 Committe CoChair
Purdue University

Elaine Scott
Bioengineering Division
Technical Program Representative, IMECE2002
Virginia Tech

CONTENTS

MEASUREMENT OF THERMAL AND MECHANICAL PROPERTIES OF BIOMATERIALS

METRICS OF THERMAL INJURY

HEAT TRANSFER IN MEDICINE AND BIOLOGY: ANALYSIS AND APPLICATIONS

Proceedings of IMECE2002
ASME International Mechanical Engineering Congress & Exposition
November 17-22, 2002, New Orleans, Louisiana

IMECE2002-33657

MEASUREMENT OF THE SHEAR MODULUS OF TISSUE-LIKE MATERIALS USING MAGNETIC RESONANCE ELASTOGRAPHY

Peter A. Hardy, University of Kentucky, Center for Biomedical Engineering, 36 MRISC, Lexington, Kentucky, 40536-0098, USA Tel: (859) 323-4723

Bao Zhang, University of Maryland, Dept. Diagnosic Radiology, Baltimore, Maryland, 21201, USA, Tel: (410) 328-8314

Dayong Gao, University of Kentucky, Dept Mechanical Engineering, 514ACRMS, Lexington, Kentucky, 40536-0108, USA, Tel: (859) 257-6043

Richard J. Mitchell, University of Kentucky, Dept Restorative Dentistry, D641 Dental Sciences Bldg, Lexington, Kentucky, 40536-0297, USA, Tel: (859) 323-5495

ABSTRACT

We developed a technology to measure the shear modulus of tissue-like materials using Magnetic Resonance (MR) imaging. In this technique, called MR elastography or MRE, the sample is vibrated at 100-300 Hz while MR images are made. The shear wave generated by the vibration will propagate into the material at a speed proportional to the square root of the shear modulus. We compared the shear modulus of gel samples measured in this way with the modulus derived from a static compression test and assuming the material is linearly elastic. The results show good agreement of the two methods provided the shear modulus of the material is below approximately 10 kPa. Differences in the two measurements are explainable by differences in the compression rate of the two techniques.

INTRODUCTION

In several instances of human disease there is manifestation of the disease as an altered mechanical property of the affected tissue. For example, physicians rely upon palpation of the breast or the prostate to detect an abnormality. The different tissue composition of a cancer, in these instances, is manifest as a stiffer material.[[1]] In many instances, however, the site of the disease is not accessible for manual palpation. Even in the case of breast and prostate cancer the non-invasive measurement of the mechanical properties of a region suspected to be abnormal would greatly facilitate patient evaluation and prevent unnecessary biopsies.

Several investigators have developed Magnetic Resonance (MR) imaging techniques to measure the characteristics of shear waves propagating into tissue.[2],[3] However, the relationship between the shear modulus measured from the MR images and other, more conventional, measurements of mechanical properties has not been clear. This work compared the shear modulus measured on MR images of tissue-like gels to the measurement derived from a static compression test using a materials testing machine.

NOMENCLATURE

MR: Magnetic Resonance
MRE: Magnetic Resonance Elastography
μ: shear modulus
E: Young's modulus
AFG: Arbitrary Function Generator
ν: Poisson's ratio

METHODS

Measurements were made on two types of gel materials: agarose (Type I, 9012-36-6; Sigma) and Gelatin (type A, EC No. 232-554-6, Porcine skin, 300 Bloom; Sigma) in varying concentrations to span the range of mechanical properties encountered in soft tissues.[4] Samples were prepared fresh for each experiment by mixing dry powder in 80°C distilled water and allowed to cool in a 4°C refrigerator. The samples were prepared in 250 ml beakers and thus were cylindrical in shape with dimensions of approximately 40 mm high by 60 mm in diameter. Identical samples prepared from a single batch were used for MR imaging measurements and for static compression measurements.

Shear wave generation To measure the shear modulus using MR imaging we constructed a device to induce shear waves into the top of a vertically-standing cylindrical sample. The vibration was induced by a flat Plexiglas disc in contact with the top of the gel sample. The disc was connected via an 80 cm long rod to a multi-turn 10 cm diameter coil mounted on a freely rotating axel. The coil was electrically connected to a stereo audio amplified (Onkyo) which amplified a sinusoidal signal produced by an arbitrary function generator (AFG) (Tektronix, AFG320). The AFG was programmed to produce a burst of sinusoidal waves upon receipt of a trigger pulse produced by the MR imager. In this way the vibration, which was transient, was synchronized with the acquisition of the MR image. Different frequencies of vibration were used ranging from 50 to 250 Hz. A schematic diagram of the experimental apparatus is shown in figure 1.

MR imaging A specially-designed MR imaging sequence was developed to be sensitive to motion in one direction. This was accomplished by modifying a spin echo pulse sequence to include two sinusoidal lobes of a motion sensitizing gradient in the frequency encoding direction. The frequency of the motion sensitizing gradient was set equal to the motion of the vibration established by the AFG.

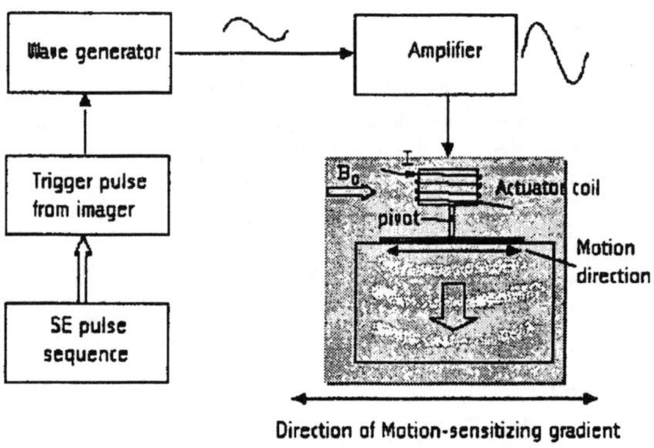

Figure 1. The configuration of the experimental apparatus used to measure the propagation of shear waves into the gel samples.

The motion encoded spin echo (SE) pulse sequence is shown in figure 2. In this figure note that the motion encoding gradients, whose shape is shown at the top of the figure, may be placed into any of the three gradients, G_{ss}, G_{ro}, or G_{pe} as shown by the shaded blocks. In our case the motion encoding gradients were always placed in the frequency encoding direction (G_{ro}). Typically we encoded motion in the X direction, i.e. left-to-right. However, we occasionally could swap the frequency and phase encoding gradients to encode motion in the Y direction. Placing the motion encoding gradients in the other axes, such as G_{pe} or G_{ss}, would encode motion in alternate directions such as Y or Z instead of X for the same image orientation. All imaging was conducted on a Siemens Vision, whole body MR imager operating at 1.5Tesla. Typically the repetition time (TR) was 150 ms and the echo time (TE) was 15 ms.

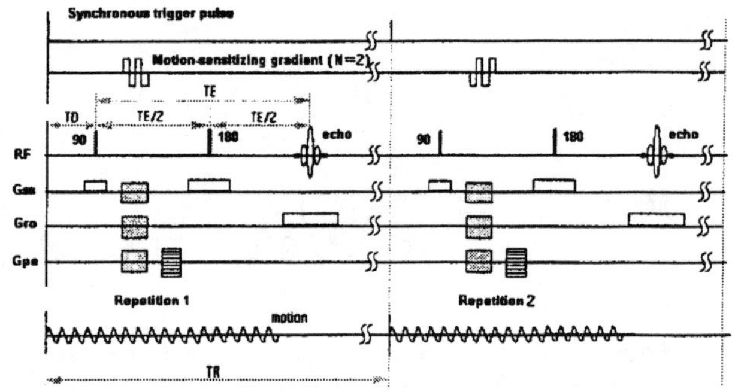

Figure 2. Diagram of the pulse sequence used to encode the motion induced by the propagation of the shear waves.

Two acquisitions with the motion encoding reversed between the two, as shown in figure 2, were employed to increase the sensitivity of the technique and to eliminate phase changes due to all other contributions other than motion. An arbitrary delay (TD) can be programmed into the sequence so that the acquisition of the image occurs at a delayed time relative to the start of the vibration shown in the bottom trace of figure 2. This delay was altered to examine the dependence of the shear motion on the duration of the vibration. From this sequence two images are produced, a magnitude image showing the location of the gel and a phase image which encoded the velocity of the motion as

$$\Phi = G_{me} a \gamma \tau \cos(\theta) \tag{1}$$

where θ is the phase difference between the motion encoding gradient (G_{me}) and the shear wave with amplitude a. The period of the encoding gradient is τ and the proton gyromagnetic ratio ($2\pi \cdot 42.577 \times 10^6$ rad.$s^{-1}T^{-1}$) is γ.

Compression tests Identical gel samples for the compression tests were prepared at the same time as those used in the MRE tests. These samples were tested on a single screw Instron Materials Testing Machine. The samples were compressed between parallel plates constructed from aluminium. The compression rate was varied but was typically 0.1 mm/sec. The force for a given compression was recorded using a chart recorder. From the slope of the force versus strain graphs the Young's modulus was estimated as:

$$E = \frac{d\sigma}{d\varepsilon} \tag{2}$$

where σ is the stress and ε is the strain in the sample. Assuming that the material is isotropic and linearly elastic, the shear modulus (μ) can be derived from E according to Eq. (3).

$$\mu = \frac{E}{2(1+\upsilon)} \quad (3)$$

where, for gels which are incompressible we have assumed the Poisson ratio ν is 0.5 simplifying Eq. (3) to $\mu = \frac{1}{3}E$.

From the MR images the shear modulus is derived by measuring the wave speed, c, either by measuring the wavelength as $c=f\lambda$ or by estimating the transit time by adjusting the delay time TD. The shear modulus is related to c as:

$$\mu = \rho c^2 \quad (4)$$

where ρ is the density of the material. [5]

RESULTS

Figure 3 shows the magnitude and phase images of a sample of gelatin produced while the sample was vibrated at 100Hz. The actuator is positioned at the top of the sample and the wave propagates from the top to the bottom of the cylinder. In the case of Figure 3, the direction of motion encoding, i.e. the X direction, is left to right which is the same direction as the amplitude of vibration. Figure 4 demonstrates the appearance of a sample when the motion encoding gradient is oriented (top-to-bottom) perpendicular to the direction of motion (left-to-right).

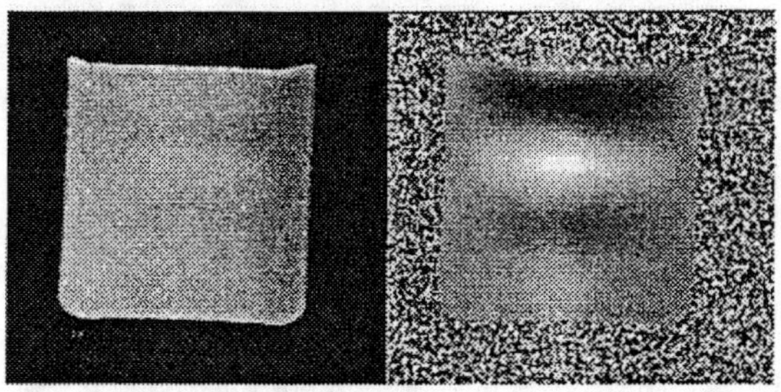

Figure 3. Magnitude (left) and phase (right) images of the propagation of a 100 Hz shear wave through a 8% gelatin sample. The actuator inducing the shear is at the top of the sample. The direction of propagation is top-to-bottom.

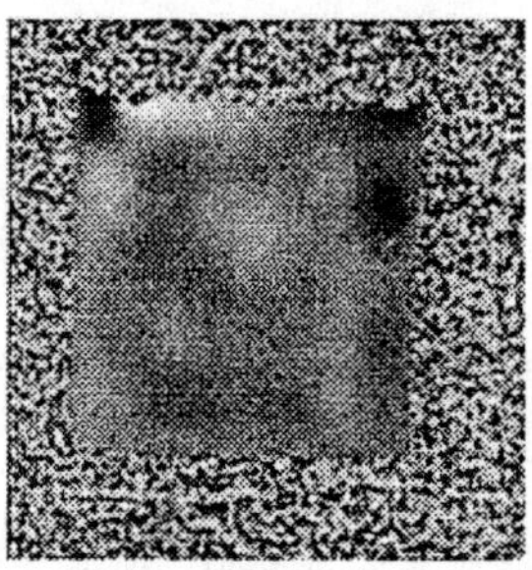

Figure 4. Phase image of a sample of gelatin produced with the motion encoding gradient perpendicular to the direction of motion.

Static measurement of λ. We estimated the wavelength, λ, of the wave in the phase image in two ways. First, by fitting the profile across the phase image through the sample to Eq. (5).

$$\Phi(x) = Ae^{-\alpha x}\sin(2\pi\frac{x}{\lambda}+\delta) \quad (5)$$

where the parameter α defines the attenuation of the wave as it progresses into the sample and λ is the wavelength of the wave. Equation 5 was fit to the phase image profiles using a Levenburg-Marquart algorithm provided in IDL to estimate λ.[6] The error in λ estimated from this fit was typically much smaller than the standard deviation derived from the collection of λ estimated from the selection of the profile in different locations on the image. Because of this we elected to estimate the wavelength directly by an electronic ruler to measure the distance between one or more crests or troughs on the phase image. The error in estimating λ in this case was related to the resolution of the image.

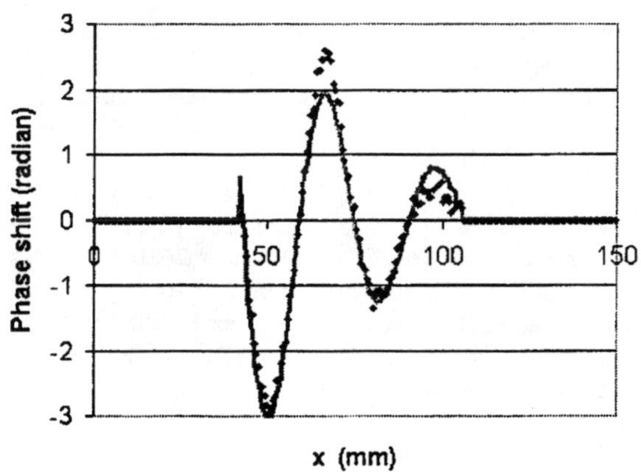

Figure 4. Profile of phase and fit of Eq. (5) to data for 100 Hz wave propagating through 2% gelatin sample.

Dynamic measurement of λ. We also measured the wave speed by making several images where we progressively altered TD from 0 to 60 ms in steps of 5 ms and monitored the movement of the wave front. An example of images generated with this method is shown in figure 5. A comparison of the wave speeds measured in one sample with the two measurements as a function of the excitation frequency is shown in figure 6. The two measurements agree over a broad range of excitation frequencies but diverge at higher frequencies. The exact reason for this divergence is not clear although we suspected that the wavelength measurement was distorted through interference at higher frequencies rather than the sample displaying a true dispersive behavior such as would be explained by a Voigt model.[7] We were unable to obtain images at excitation frequencies higher than 250 Hz to explore this phenomena further.

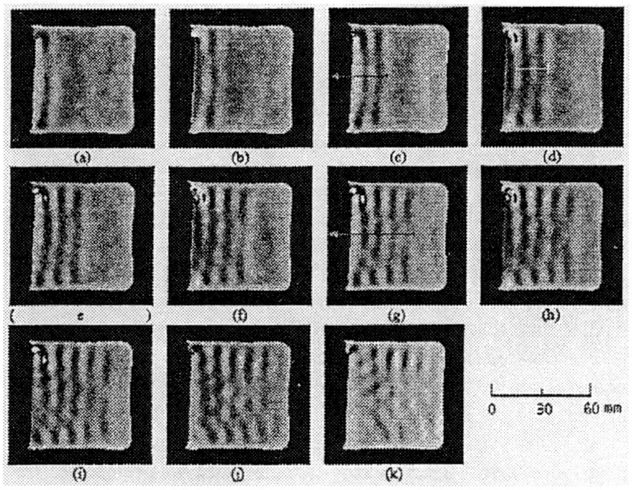

Figure 5, phases images of a sample of gelatin taken at different delay times (TD) from 0 to 60 ms in 5 ms intervals. Note the alteration of the wave front and the generation of interference patterns as the wave moves down into the sample and reflections arise from the bottom of the beaker containing the gelatin.

The shear modulus measured in the gelatin samples with MRE varied linearly with the concentration of gelatin. The shear modulus increased from 1.06±0.12 to 14.26±0.46 kPa as the concentration of gelatin increased from 2 to 10% by weight. In agarose phantoms the shear modulus estimated by MRE increased from 2.67±0.19 to 45.9±0.8 kPa as the agarose concentration increased from 0.5 to 2% by weight. This was a non-linear increase.

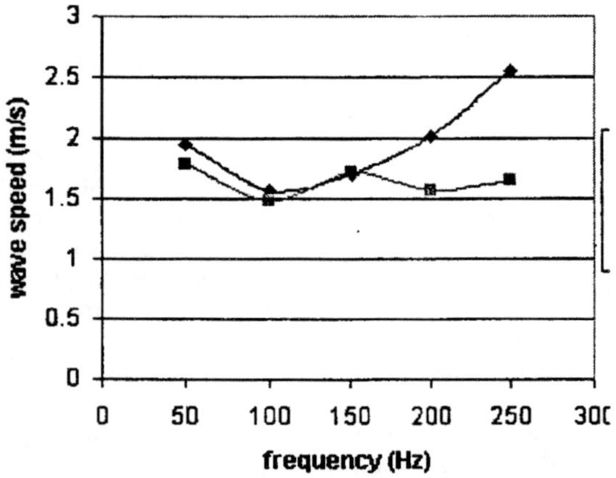

Figure 6. Comparison of the wave speed estimated from the wavelength measurement (upper curve, filled diamonds) and the migration of the wave front measurement in a 0.5% agarose sample.

We also measured the Young's modulus and calculated the shear modulus according to Eq. (3) from the static compression tests. These measurements confirmed the linear increase in shear modulus with gelatin and agarose concentration.

Figure 7 demonstrates the comparison of the measurement of the shear modulus using MRE with that from the static compression test.

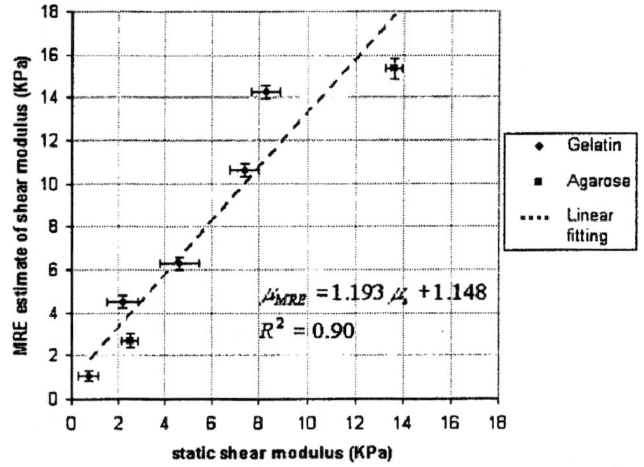

Figure 7, comparison of μ derived from MRE experiments and μ derived from static compression tests on two types of samples. The dotted line is a linear fit to the data. Error bars represent one standard deviation.

DISCUSSION

The results of figure 7 demonstrate that the two methods of measuring the shear modulus are proportional but are not identical. The results were highly correlated (R^2=0.90) however, the slope was 1.19±0.18. The difference between the two measurements, we suspected, arose from the different load rates used in the two measurements. In the case of the Instron measurement the typical load rate was 0.1 mm/s while for the MRE experiment, operating at 100Hz and with an amplitude of vibration of 140 μm which we measured directly and estimated from the measured phase change, the calculated load rate is approximately 56 mm/s a factor of 560 times larger. To investigate the dependence of the Young's modulus on the load rate we remeasured the modulus on one sample, (2% agarose) using increasing load rates between 0.025 and 0.50 mm/s. The results show a 10% increase in Young's modulus with load rate in this range indicating the material is viscoelastic. The viscoelastic nature of the material was also evident from the creep in the sample once the force was held constant and the gradual exudation of water from a sample under compression.

CONCLUSIONS

The shear modulus estimates from MRE is approximately equal to that estimated from static compression tests only for samples with shear modulus less than approximately 10 kPa. However, several different tissue types including fat, fibroglandular tissue and breast tumor have shear modulus in this range. However, other tissue types, such as muscle[8], articular cartilage[9,10] can have much higher elastic moduli and they are strongly viscoelastic. For these tissues the MRE estimate will likely differ substantially from more conventional methods for measuring shear modulus.

ACKNOWLEDGMENTS

This work was supported by a grant (#RG-98-0535) from the Whitaker Foundation.

REFERENCES

[1] Samani A, Bishop J, Sciarretta J, Plewes D, Breast magnetic resonance elastography: a new reconstruction technique using MRI derived constraints. Proc. Intl. Soc Magn reson Med. 8 (2000), p2174.

[2] Muthupillai R, Lomas D, Rossman P, Greenleaf JF, et al. Magnetic Resonance elastography by direct visualization of propagating acoustic strain waves, Science 269:1854-1857, 1995.

[3] Plewes DB, Bishop J, Samani A, Sciarretta J, Visualization and quantification of breast cancer biomechanical properties with magnetic resonance elastography, Phys. Med. Biol 45:1591-1610, 2000.

[4] Mitchell MD, Kundel HI, Axel L, Joseph P, Agarose as a tissue equivalent phantom material for NMR imaging. Mag Reson Imaging 4:263-266, 1986.

[5] Ugural AC, Fenster SK, Advanced strength and applied Elasticity. 3rd edition, pp 1-90, Prentice – Hall, New Jersey, 1995.

[6] IDL produced by Research Systems Incorporated, Colorado.

[7] Kolsky H, Stress Waves in Solids, Dover Publishing, New York, 107-109, 1963.

[8] Wu T, Felmlee JP, Greenleaf JF, Riederer SJ, Ehman RL, Assessment of thermal tissue ablation with MR elastography, Proc. Intl Soc Magn Reson Med 8 (2000), p46.

[9] Hoch DH, Grodzinsky AJ, Koob GJ, Albert ML, Eyre DR, Early changes in material properties of rabbit articular cartilage after meniscectomy, J Orthop Research I:4-12, 1983.

[10] Kruse SA, Smith JA, Lawrence AJ, et al, Tissue characterization using magnetic resonance elastography: shear modulus and shear viscosity. Proc Int Soc Magn Reson Med 8 (2000), p1976.

Proceedings of IMECE2002
ASME International Mechanical Engineering Congress & Exposition
November 17-22, 2002, New Orleans, Louisiana

IMECE2002-33664

EFFECT OF THERMAL PROPERTIES ON HEAT TRANSFER IN CRYOPRESERVATION AND CRYOSURGERY

Bumsoo Han and John C. Bischof
Department of Mechanical Engineering
University of Minnesota, Minneapolis, MN55455

ABSTRACT

Biological materials in both cryopreservation and cryosurgery are composed of various chemicals and experience a wide range of temperature change. Therefore, their thermal properties including specific heat, latent heat (including water/ice and eutectic phase change) and thermal conductivity are expected to change significantly during freezing/thawing. The effects of thermal properties on heat transfer in cryopreservation/cryosurgery were studied experimentally and numerically. Thermal properties of various biological aqueous solutions were measured over a wide temperature range (-150 ~ 30°C). To estimate the effect of thermal property changes on the heat transfer, numerical simulations of both cryopreservation (cooled from outside) and cryosurgery (cooled from inside) geometries were performed with constant and temperature-dependent properties. The results show that the constant-property case significantly under-predicts the heat transfer over the temperature-dependent-property case regardless of the geometry.

INTRODUCTION

To improve the quality of cryopreservation and cryosurgery protocols, accurate and reliable prediction of heat transfer within biological systems is vital. However, the ability to predict the heat transfer is limited due to an insufficient thermal property database of biological materials including specific heat, latent heat and thermal conductivity [1]. Even though biological materials in both cryopreservation and cryosurgery experience a wide range of temperature change (from body temperature to -160°C), most thermal property data are only available within a relatively narrow temperature range (from body temperature to -40°C) [2]. Additionally, since most biological materials are composed of various chemicals and are usually loaded with additives such as cryoprotective agents (CPA) in cryopreservation, their thermal properties are expected to change significantly with CPA composition and concentration.

Thermal properties of various biological systems have been reviewed and summarized in several references [3-7]. However, as mentioned above, most data are only available within a relatively narrow temperature range (from body temperature to -40°C). Among the thermal properties under consideration, thermal conductivity has been most extensively studied due to its thermodynamical importance. Cooper and Trezek [8] measured the thermal conductivity of various media by measuring step temperature change of a thermocouple probe. A transient self-heated thermistor technique was introduced to measure thermal conductivity and diffusivity of biological solutions in [9]. More studies were performed in biological solutions, tissue phantoms and in-vitro/in-vivo tissues [10, 11] with modified transient self-heated thermistor techniques. Further developments of thermal conductivity and diffusivity measurement techniques using thermistors were achieved in [12]. As an alternative technique, a differential scanning calorimeter (DSC) has been used to measure thermal conductivity [13-15], but few results are reported for biological systems by this method.

It has been experimentally shown that the latent heat of biological systems is quite different from that of pure water. Fennema et al. [3] observed that the magnitude of the latent heat of various foods was less than that of pure water, and suggested that the differences could be correlated with their water content. Devireddy et al. [16] measured the latent heat of various biological solutions using a DSC. They also found a reduction in the magnitude of latent heat, but concluded that the reduction did not correlate with water content only. Contrary to above two thermal properties – thermal conductivity and latent heat, available specific heat data are relatively limited and are usually approximated to a simple weighted average value of components in the system [6].

Numerical models have been useful tools to predict heat transfer in cryopreservation and cryosurgery. Numerical models of the freezing process in cryopreservation/cryosurgery have been developed using different algorithms including front-tracking methods [17, 18], enthalpy methods [19-22], and explicit source term methods [23, 24]. However, since most effort was focused on the development of solution algorithms, thermal properties such as specific heat and thermal conductivity were assumed constants in both the solid and liquid regions. Temperature dependence of the thermal properties was considered in several studies [22, 24]. The difference in latent heat release of biological systems from that of water was first considered by Hayes at el. [21, 25]. They implemented the latent heat release from the water-NaCl binary phase diagram into their numerical model and performed numerical simulations of freezing bulk systems. Their results showed that the temperature history within the systems was significantly affected by the latent heat release pattern. Recently, a numerical study separating latent heat into latent heat of extracellular and intracellular water was performed [26]. Even though the above advances have been made, insufficient thermal property data still limit the accuracy and predictive ability of numerical models.

In the present study, thermal properties including specific heat, latent heat and thermal conductivity of several biological solutions were measured. The biological solutions studied were NaCl-water (1x (isotonic), 5x and 10x), phosphate buffered saline (PBS) (1x (isotonic), 5x and 10x), and PBS+glycerol (0.1M, 1M and 2M glycerol + 1xPBS). To estimate the effect of thermal property changes on the heat transfer, numerical simulations on both cryopreservation (cooled from outside) and cryosurgery (cooled from inside) geometries were performed with three different assumptions on thermal properties – 1) constant thermal properties in each phase and latent heat release at single temperature; 2) temperature dependent thermal properties and latent heat release at a single temperature; and 3) temperature dependent thermal properties and latent heat release of 1xNaCl-water including both water/ice and eutectic phase change. Comparison among the numerical simulation results illustrates the significance of thermal properties in cryobiological applications.

EXPERIMENTAL AND NUMERICAL METHODS

Two independent experimental setups were developed and used to measure the thermal properties – 1) Differential Scanning Calorimeter (Perkin-Elmer, DSC-Pyris 1) for specific heat and latent heat and 2) a pulse-decay setup for thermal conductivity. For the measurements, following biological solutions were prepared – 1) 1x (isotonic or 0.15M), 5x (0.75M) and 10x (1.5M) NaCl-water solution; 2) 1x, 5x and 10x phosphate buffered saline (PBS, Life Technology, D-PBS) solutions; and 3) PBS+glycerol solutions (0.1M, 1M and 2M glycerol + 1xPBS). The concentrations of the solutions were checked with an osmometer (Precision Systems, Micro-Osmometer), and the uncertainty in the concentration was less than ± 3% of its nominal value.

Before the specific heat and latent heat measurement, the temperature scale of the DSC was carefully calibrated using two different transition temperatures of cyclohexane (-85.8°C and 6.4°C). The heat flow scale of DSC was calibrated against the heat of fusion of pure water (335J/g) during thawing at 5°C/min. The DSC experiments were conducted with 6~9mg of each solution in a standard aluminum sample pan. The sample was pre-nucleated by the method developed in [16]. After the pre-nucleation, the sample was cooled to -150°C and warmed up to 50°C at 5°C/min. To identify the difference between the specific heat measurement during cooling and warming, the results of specific heat measurements were performed both during cooling and warming, and compared. The comparison confirmed that differences were negligible. All the data presented in this paper were measured during warming. For impure solutions including NaCl-water solution and PBS solution, the specific heat measurements were only performed where no phase change occurred to exclude latent heat from the evaluation of specific heat.

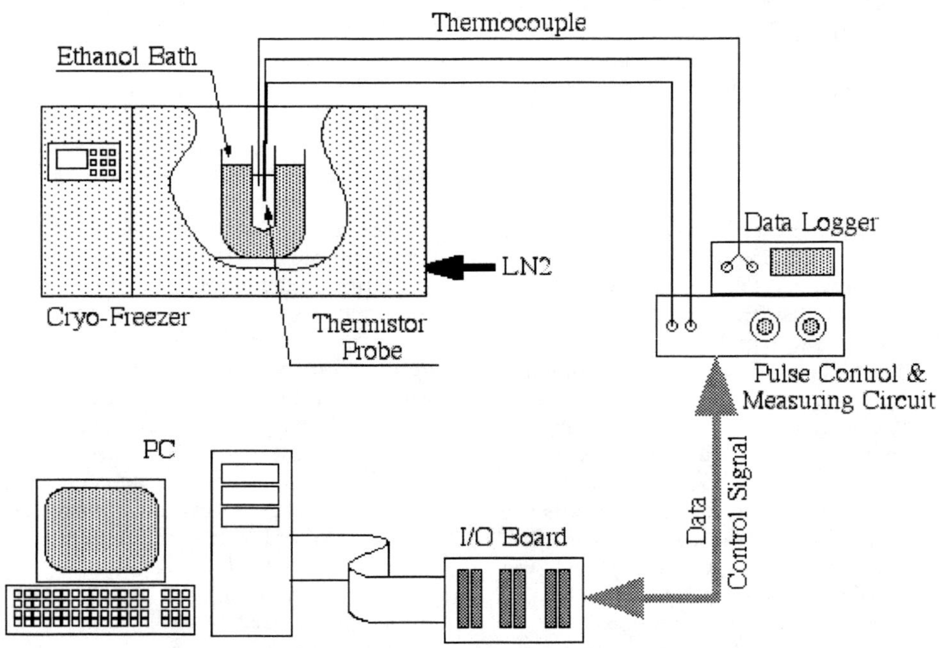

Fig.1. Schematic diagram of experimental setup for thermal conductivity measurements

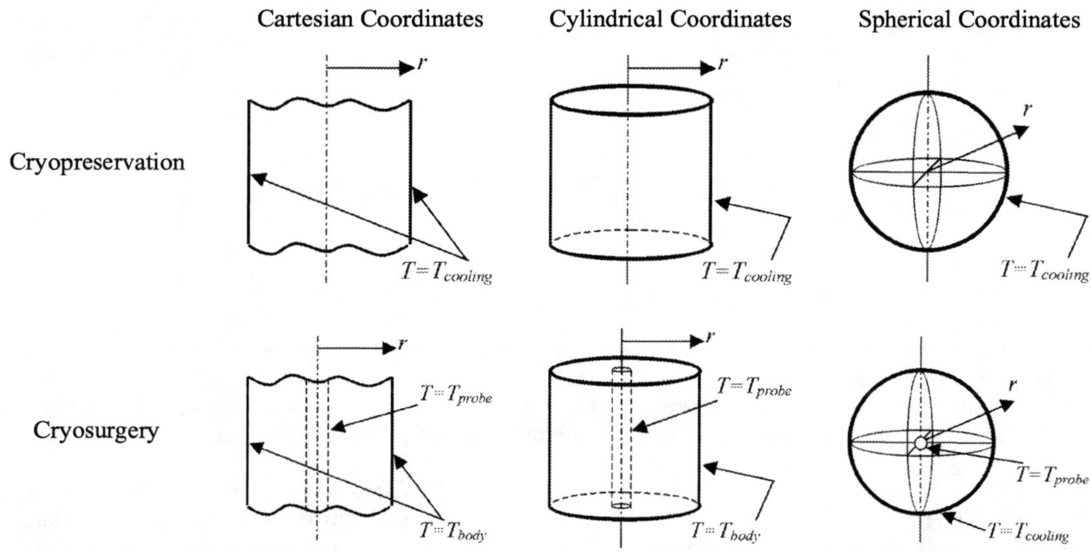

Fig. 2. Schematic diagrams of calculation geometry and boundary conditions

For the latent heat measurements, the phase change temperatures of the sample were estimated through a preliminary cooling/warming run within the same temperature range (-150°C~50°C). After the phase change temperature range was estimated, the heat release from the sample was measured through a main run at a slower warming rate (~1°C/min) to ensure approximate thermodynamic equilibrium. Therefore, the phase change temperatures and the magnitudes of the latent heat measured were comparable to the published phase diagram data [27].

In order to measure the thermal conductivity, an experimental setup was designed and built, which employed a pulse-decay method [10]. The pulse-decay method is a technique to evaluate thermal conductivity by measuring the temperature decay of a thermistor which is heated by a known electric power. A schematic diagram of the thermal conductivity setup is shown in Fig. 1. A thermistor (Thermometrics, P60BA120K) with small bead diameter (~1mm) was submerged in the sample which is placed inside of an ethanol isothermal bath to minimize the temperature fluctuation during the measurement. The temperature of the sample was controlled using a cryo-freezer (Planer, Kryo10 series III) from -50°C~ 20°C, and monitored with a T-type thermocouple connected to a data logger (Fluke, Hydra Data logger). The pulse control and measuring circuit was developed based on the one in [10]. The duration of the thermistor heating was controlled by a personal computer through an I/O board (National Instrument, PCI-6035); the temperature decay data of the thermistor was also gathered through the I/O board.

To illustrate the effect of thermal property variations on heat transfer in cryopreservation and cryosurgery applications, several one-dimensional unsteady heat conduction problems with phase change were constructed. Schematic diagrams of each problem and corresponding boundary conditions are shown in Fig. 2. All the systems had a uniform temperature distribution initially (T_{body} = 37°C), and were cooled from either outside (cryopreservation) or inside (cryosurgery) to -160°C. The governing equation can be written in terms of enthalpy and temperature for a general one-dimensional coordinate system as follows:

$$\frac{\partial(\rho h)}{\partial t} = \frac{1}{r^n}\frac{\partial}{\partial r}\left(r^n k \frac{\partial T}{\partial r}\right) \qquad (1)$$

where ρ is density, h is specific enthalpy, t is time, r is a spatial coordinate, k is thermal conductivity, and T is temperature. n is a geometric parameter – 0 for Cartesian, 1 for cylindrical and 2 for spherical coordinates. The enthalpy h can easily be related to the temperature when the specific heat of a system is known [28, 29]. The governing equation was numerically solved by an explicit enthalpy method. The following three cases were simulated and compared to show the effect of thermal property variations on heat transfer in cryopreservation/cryosurgery applications.

1. The case with constant thermal properties in each phase and latent heat release at single temperature.

2. The case with temperature dependent thermal properties and latent heat release at single temperature.

3. The case with temperature dependent thermal properties and latent heat release of 1xNaCl-water including both water/ice and eutectic phase change.

The thermal properties used for numerical simulations are summarized in Table 1.

Table 1. Thermal properties used in numerical simulations

	Thermal conductivity (W/m•K)		Specific heat (J/g•K)		Latent heat	Density (kg/m³)	
	Solid	Liquid	Solid	Liquid		Solid	Liquid
Case 1	2.0	0.552	2100	4200	335J/g at 0°C		
Case 2	$2.24+0.005975(273-T)^{1.156}$	0.552	7.16T+138	4200	335J/g at 0°C	921	999
Case 3	$2.24+0.005975(273-T)^{1.156}$	0.552	7.16T+138	4200	299.4(0.001132-0.53/T) J/g from 272.65K to 252K		

Note: Temperatures in Kelvin
Temperature dependent thermal conductivity and specific heat expressions were obtained from [29].
Latent heat release patterns of Case3 was obtained by DSC data from the present study.

RESULTS AND DISCUSSION

The measured specific heat data of pure water, NaCl-water, PBS solutions and PBS+glycerol with different concentrations are shown in Fig. 3. In the liquid phase, the specific heat is reduced with the concentration of solutes, and the amount of reduction can be easily estimated by weighted averaging of the specific heat of each component in the samples. In the solid phase, however, no significant difference is noticed in NaCl-water and PBS samples with respect to pure water, though PBS show slightly less specific heat than NaCl-water at the same concentration. When glycerol is added to 1xPBS, a significant increase of the specific heat is observed from -80°C to -30°C, which is proportional to the concentration of glycerol added. This increase in the specific heat cannot be explained by weighted averaging of each component's specific heat. One possible explanation of the observed increase can be made by the phase change characteristics of PBS+glycerol solutions. Adding glycerol into PBS solutions disperses the eutectic phase change during freezing [30]. As a result, the sample of PBS+glycerol solution below -30°C is a mixture of ice crystals and unfrozen concentrated solutions. Therefore, their thermal properties should be affected by not only the composition of samples, but also the phase of each component at a given temperature. This may lead to a significant increase in the specific heat over the temperature range because the specific heat of the liquid phase is much larger than that of the solid phase.

Latent heat release patterns of the solutions during thawing are plotted versus temperature in Fig. 4. Since NaCl-water solution is a simple binary mixture, some amount of heat release associated with eutectic phase change is observed around the eutectic temperature -22°C, regardless of the concentration. For isotonic NaCl-water, about 3% of the total latent heat of melting is released due to the eutectic melting. The heat release due to the eutectic phase change increased to about 15% for 5xNaCl-water solution and 30% for 10xNaCl-water solution. The eutectic phase changes are also observed in PBS solutions, even though they are not binary mixtures. However, contrary to NaCl-water, the heat release associated with eutectic transition occurs from -27°C to -22°C, which is expected due to other electrolytes in the solutions. Even though the eutectic heat release patterns of PBS solutions are wider than those of NaCl-water, the percentage of the eutectic heat release is very similar to that of NaCl-water solutions at the same concentration. As glycerol was added to 1xPBS solution, the eutectic heat release disappears in all concentrations studied (≥ 0.1M). A more detailed study on the latent heat of eutectic phase change and the effect of adding glycerol can be found in [30]. The magnitudes of latent heat with the phase change temperatures are summarized at Table 2.

In Fig. 5, the results of thermal conductivity measurements are presented. To verify the validity of the present measurement, thermal conductivity of pure water is measured and compared with the literature [5]. As shown, the measured values agree very well with the literature values. The thermal conductivity of NaCl-water and PBS solutions increase due to high thermal conductivity of solutes including NaCl. Therefore, thermal conductivity generally increases with the concentration of solute, and the effect of the concentration is much larger in solid phase than in liquid phase. In solid phase, thermal conductivity of PBS sample is larger than that of NaCl-water at the same concentration. Further study is required to extend the temperature range and to investigate the effect of adding glycerol or other types of CPA on thermal conductivity.

The numerical simulation results of the cryopreservation cases are presented in Fig. 6. The freezing interface propagation shows that the case with constant property assumption significantly under-predicts the propagation of the freezing front in all three different coordinates. For example, in spherical coordinates, constant property calculation predicts the position of the freezing interface as 3.8cm in comparison to 1.8cm of temperature dependent thermal property at 1200sec. This is about 110% under-prediction of the freezing phenomena, and causes a significant difference in the prediction of local cooling rates as well.

Corresponding temperature histories in cylindrical coordinates at several locations are shown in Fig. 6 (b), and they are also severely under-predicted. The numerical results for the cryosurgery cases are shown in Fig. 7. Even though the magnitude of under-prediction is lessened due to the geometric

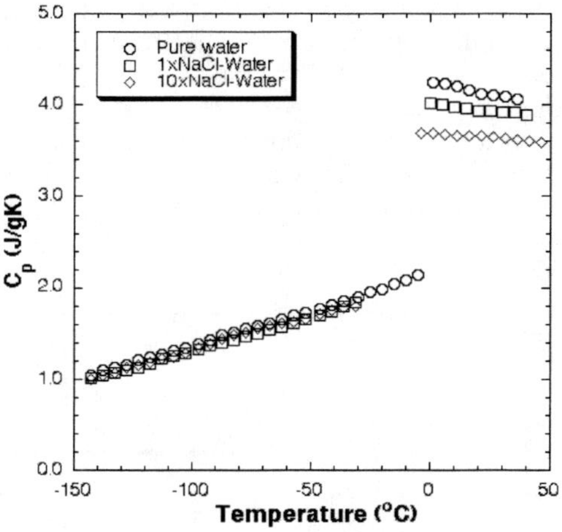

(a) NaCl-water solutions

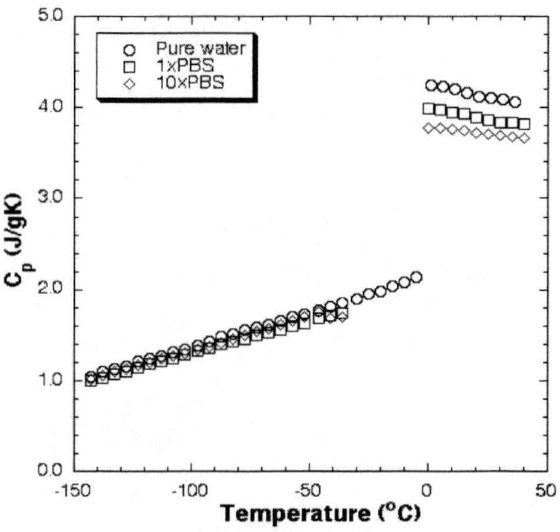

(b) PBS solutions

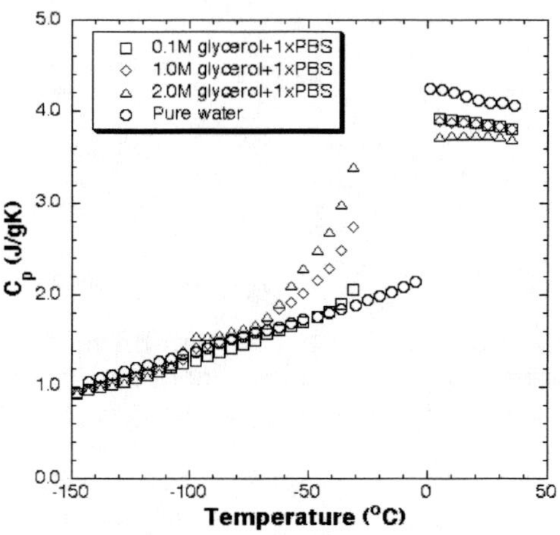

(c) Glycerol+1xPBS solutions

Fig. 3. Effect of concentration and composition of biological solutions on specific heat

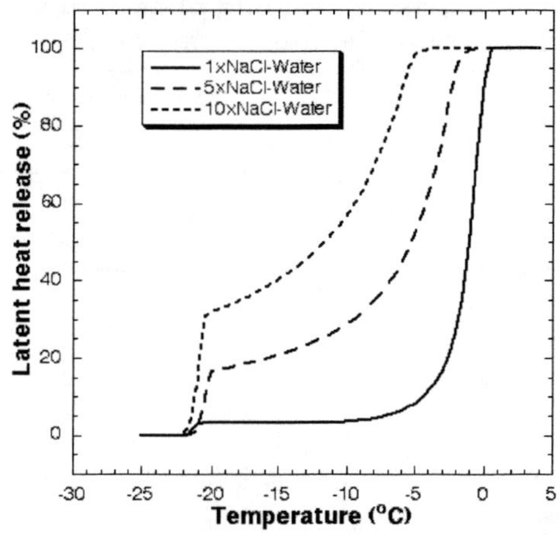

(a) NaCl-water solutions

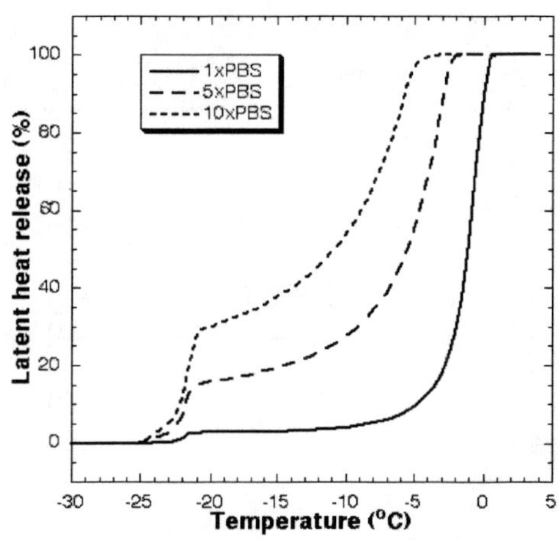

(b) PBS solutions

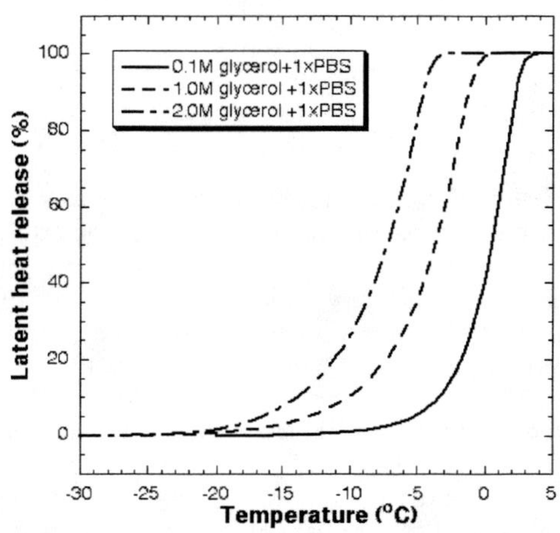

(c) Glycerol+1xPBS solutions

Fig. 4. Effect of concentration and composition of biological solutions on latent heat release

Table 2. Latent heat and phase change temperature of biological solutions

Solutions	Latent Heat (J/g)	Phase Change Temperature (°C)	
		Ice-melting	Eutectic-melting
1xNaCl-water	299.4	-0.5	-22.8
5xNaCl-water	291.5	-2.6	-22.5
10xNaCl-water	279.3	-6.0	-23.0
1xPBS	298.3	-0.5	-25.7
5xPBS	287.9	-3.0	-26.3
10xPBS	273.7	-5.9	-26.4
0.1M Gly.+1xPBS	272.7	-3.2	N/A
1.0M Gly.+1xPBS	220.9	-7.2	N/A
2.0M Gly.+1xPBS	169.7	-11.8	N/A

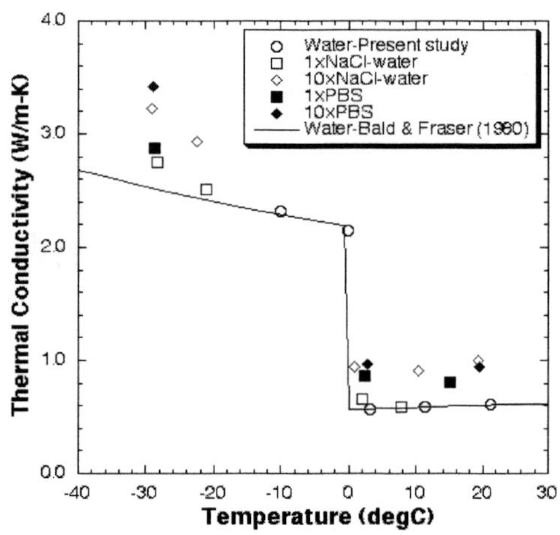

Fig. 5. Effect of concentration and composition of biological solutions on thermal conductivity

characteristics of cryosurgery (cooling from inside out), the constant thermal property case still under-predict the propagation of the freezing front. Similar results were also reported in [24].

The effect of latent heat release patterns on heat transfer in cryopreservation is investigated by comparing the freezing front propagation of cryopreservation in cylindrical coordinates. For case 3, the latent heat release data of 1xNaCl-water in Fig. 4 are curve-fitted and implemented to the numerical code. Figure 8 (a) clearly shows that the assumption of latent heat release at a single temperature also under-predict the freezing front propagation. Based on the position at 1200s, case 1 underestimates the location of the freezing front by 100% compared to case 3. In the thermal history shown in Fig. 8 (b), the degree of under-prediction becomes worse as one approaches the centerline of the domain. The temperature at the centerline at t=2000sec in case 3 is -123°C, contrary to 15°C in case 1. Local cooling rate changes due to latent heat release patterns were also reported in [25]. Therefore, information of thermal properties including specific heat, thermal conductivity and latent heat are critical to improve the accuracy of numerical models.

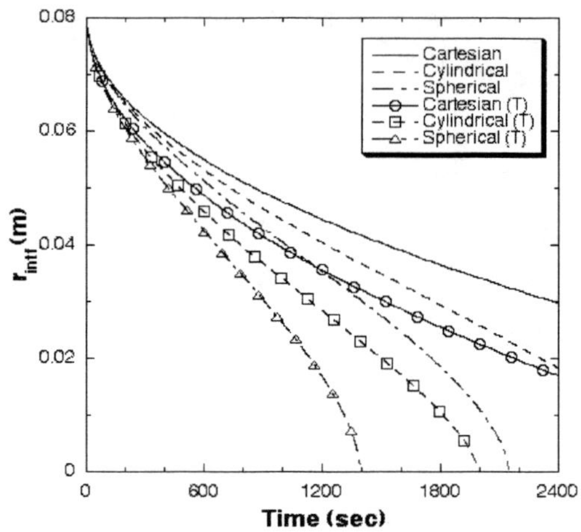

(a) Solid-liquid interface propagation

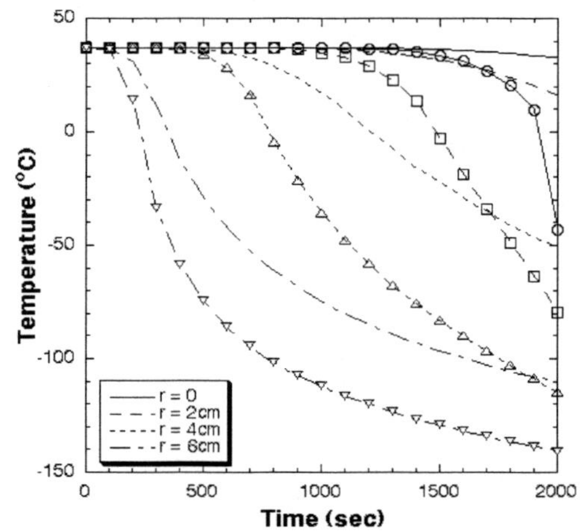

(b) Thermal history in cylindrical geometry

Fig. 6. Effect of thermal properties on heat transfer in cryosurgery (Lines without symbol are for the result of case 1, and lines with hollow symbols for the result of case 2.)

CONCLUSION

To estimate the effects of thermal properties on heat transfer in cryopreservation/cryosurgery, an experimental and numerical study was performed. Thermal properties of various biological solutions including specific, latent heat and thermal conductivity were measured with temperature, concentration

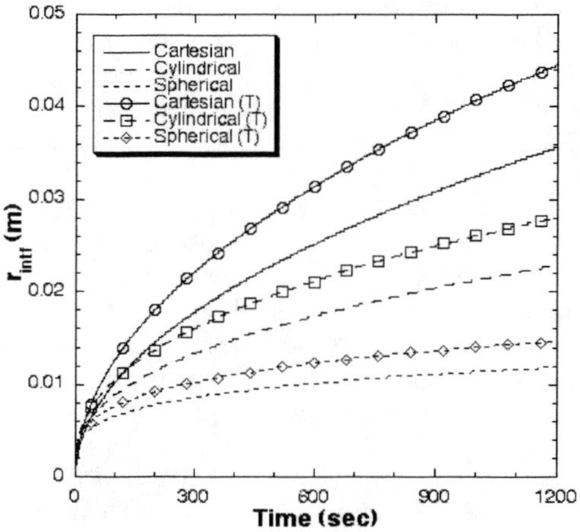

(a) Solid-liquid interface propagation

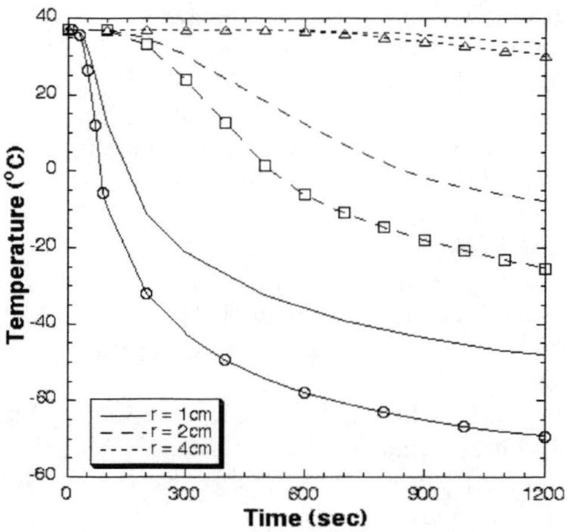

(b) Thermal history in cylindrical geometry

Fig. 7. Effect of thermal properties on heat transfer in cryosurgery (Lines without symbol are for the result of case 1, and lines with hollow symbols for the result of case 2.)

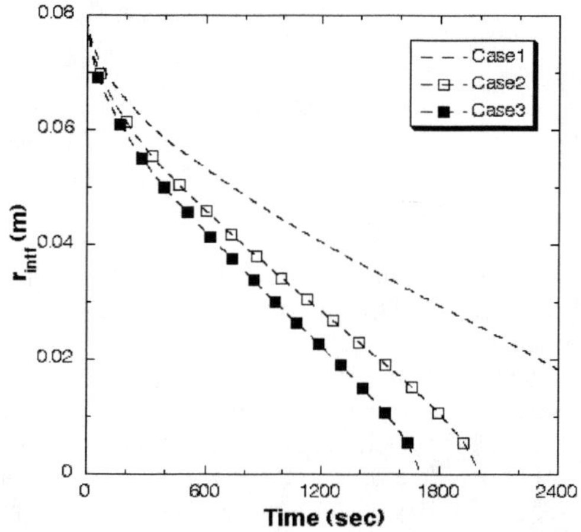

(a) Solid-liquid interface propagation

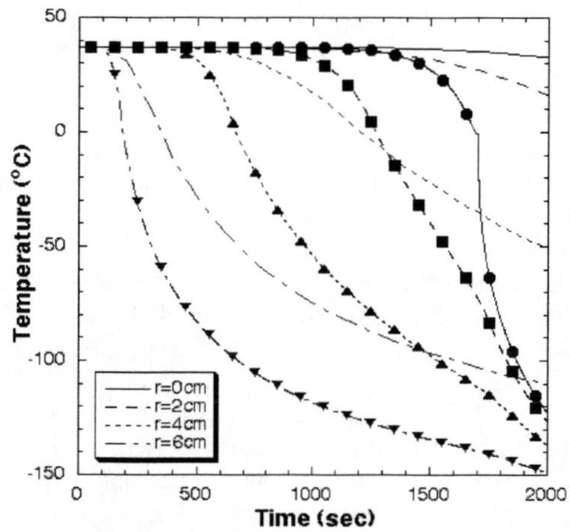

(b) Thermal history in cylindrical geometry

Fig. 8. Effect of latent heat release pattern on heat transfer in cryopreservation (Lines without symbol are for the result of case 1, lines with hollow symbols for the results of case 2, and lines with solid symbols for the result of case 3.)

and composition change. A corresponding experimental setup and procedures were also developed for the measurement of each property. Specific heat and latent heat release were measured with DSC, and thermal conductivity was obtained by a pulse-decay method. For specific heat, changes in concentration and composition of saline solutions (NaCl-water or PBS) are not from that of pure water. However, significant variation is observed when glycerol was added to saline solutions. The addition of glycerol also eliminates the eutectic phase change during freezing and thawing, and changes the latent heat release patterns. Contrary to specific heat, thermal conductivity was noticeably increased with increasing concentration of saline in both the solid and liquid phase over the temperature range studied.

To estimate the effect of thermal property change on heat transfer in cryopreservation/cryosurgery, numerical simulations were performed in Cartesian, cylindrical and spherical coordinates with the following assumptions - case 1: constant thermal properties in each phase and latent heat release at single temperature; case 2: temperature dependent thermal properties and latent heat release at single temperature; and case 3: temperature dependent thermal properties and latent heat release over a certain temperature range. The numerical results show that case 1 significantly under-predict the freezing interface propagation and temperature history at a given location in both cryopreservation and cryosurgery applications.

ACKNOWLEDGMENTS

This work was supported by a grant from the National Science Foundation (NSF-BES 9703326). The authors would like to thank them for their financial support.

REFERENCES

[1] Diller, K., 1992, "Modeling of Bioheat Transfer Processes at High and Low Temperatures," Bioengineering Heat Transfer, Y. I. Choi, eds., **22**, Academic Press, pp. 157-358.

[2] Bischof, J. C. and Han, B., 2002, "Cryogenic Heat and Mass Transfer in Biomedical Applications," Proceedings of 12th International Heat and Mass Transfer Conference, France.

[3] Fennema, O. R., Powrie, W. D., and Marth, E. H., 1973, *Low-Temperature Preservation of Foods and Living Matter*, Marcel Dekker, Inc., New York.

[4] Bowman, H. F., Cravalho, E. G., and Woods, M., 1975, "Theory, Measurement and Application of Thermal Properties of Biomaterials," Annual Review of Biophysics and Bioengineering, **4**, pp. 43-82.

[5] Bald, W. B. and Fraser, J., 1982, "Cryogenic Surgery," Reports on Progress in Physics, **45**, pp. 1381-1434.

[6] Chato, J. C., 1985, "Measurement of Thermal Properties of Biological Materials," Heat Transfer in Medicine and Biology, Shitzaer and Eberhart, eds., pp. 167-192.

[7] Diller, K. R., Valvano, J. W., and Pearce, J. A., 1999, "Bioheat Transfer," The CRC Handbook of Thermal Engineering, F. Kreith, eds., pp. 4114-4187.

[8] Cooper, T. E. and Trezek, G. J., 1972, "A Probe Technique for Determining the Thermal Conductivity of Tissue," Journal of Heat Transfer, pp. 133-140.

[9] Balasubramaniam, T. A. and Bowman, H. F., 1976, "Thermal Conductivity and Thermal Diffusivity of Biomaterials: A Simultaneous Measurement Technique."

[10] Chen, M. M., Holmes, K. R., and Rupinskas, V., 1981, "Pulse-Decay Method for Measuring the Thermal Conductivity of Living Tissues," Journal of Biomechanical Engineering, **103**, pp. 253-260.

[11] Bai, X. and Pegg, D. E., 1991, "Thermal Property Measurements on Biological at Subzero Temperatures," Journal of Biomechanical Engineering, **113**, pp. 423-429.

[12] Valvano, J. W., 1988, "Low Temperature Tissue Thermal Properties," Low Temperature Biotechnology: Emerging Applications and Engineering Contributions, J. J. McGrath and K. R. Diller, eds., **BED-Vol. 10**, New York, pp. 331-345.

[13] Chiu, J. and Fair, P. G., 1979, "Determination of Thermal Conductivity by Differential Scanning Calorimetry," Thermochimica Acta, **34**, pp. 267-273.

[14] Khanna, Y. P., Taylor, T. J., and Chomyn, G., 1988, "A New Differential Scanning Calorimetry Based Approach for the Estimation of Thermal Conductivity of Polymer Solids and Melts," Polymer Engineering and Science, **28**, pp. 1034-1041.

[15] Keating, M. Y. and McLaren, C. S., 1990, "Thermal Conductivity of Polymer Melts," Thermochimica Acta, **166**, pp. 69-76.

[16] Devireddy, R. V., Leo, P. H., Lowengrub, J. S., and Bischof, J. C., 2001, "Measurement and Numerical Analysis of Freezing in Solutions Enclosed in a Small Container," International Journal of Heat and Mass Transfer, pp. In press.

[17] Rubnisky, B. and Cravalho, E. G., 1981, "A Finite Element Method for the Solution of One-Dimensional Phase Change Problems," International Journal of Heat and Mass Transfer, **24**, pp. 1987-1989.

[18] Bischof, J. C., Bastacky, J., and Rubinsky, B., 1992, "An Analytical Study of Cryosurgery in the Lung," Journal of Biomechanical Engineering, **114**, pp. 467-472.

[19] Comini, G. and Giudice, S. D., 1976, "Thermal Aspect of Cryosurgery," Journal of Heat Transfer, pp. 543-549.

[20] Budman, H., Shitzer, A., and Giudice, S. D., 1986, "Investigation of Temperature Fields around Embedded Cryoprobes," Journal of Biomechanical Engineering, **108**, pp. 42-48.

[21] Hayes, L. J., Diller, K. R., and Chang, H., 1986, "A Robust Numerical Method for Latent Heat Release During Phase Change," Numerical Methods in Heat Transfer, J. L. S. Chen and K. Vafai, eds., **62**, ASME, New York, pp. 63-70.

[22] Rabin, Y. and Shitzer, A., 1998, "Numerical Solution of the Multidimensional Freezing Problem During Cryosurgery," Journal of biomechanical engineering, **120**, pp. 32-37.

[23] Hayes, L. J. and Diller, K. R., 1983, "Implementation of Phase Change in Numerical Models of Heat Transfer," Journal of Energy Resources Technology, **105**, pp. 431-435.

[24] Smith, D. J., Devireddy, R. V., and Bischof, J. C., 1999, "Prediction of Thermal History and Interface Propagation During Freezing in Biological Systems - Latent Heat and Temperature-Dependent Property Effects," 5th ASME/JSME Joint Thermal Engineering Conference.

[25] Hayes, L. J. and Diller, K. R., 1988, "Computational Methods for Analysis of Freezing Bulk Systems," Low Temperature Biotechnology: Emerging Applications and Engineering Contributions, J. J. McGrath and K. R. Diller, eds., **BED-Vol. 10**, New York, pp. 253-272.

[26] Devireddy, R. V., Smith, D. J., and Bischof, J. C., 2002, "Effect of Microscale Mass Transport and Phase Change on Numerical Prediction of Freezing in Biological Tissues," Journal of Heat Transfer, (in press).

[27] Hohne, G., Hemminger, W., and Flammersheim, H. J., 1996, *Differential Scanning Calorimetry: An Introduction for Practitioners*, Springer.

[28] Ozisik, M. N., 1993, *Heat Conduction*, Wiley, New York.

[29] Alexiades, V. and Solomon, A. D., 1993,

Mathematical Modeling of Melting and Freezing Processes, Hemisphere Publishing Crop., Washington.

[30] Han, B., Devireddy, R. V., and Bischof, J. C., 2002, "Phase Change Behavior of Biological Solutions," Proceedings of ASME 2002 International Mechanical Engineering Congress and Exposition, submitted for review.

Proceedings of IMECE2002
ASME International Mechanical Engineering Congress & Exposition
November 17-22, 2002, New Orleans, Louisiana

IMECE2002-33683

A TINY DETECTOR MADE OF A SELF-HEATED THERMISTOR FOR DETERMING THERMAL CONDUCTIVITY OF BIOMATERMIALS IN TEMPERATURE RANGE 233~313K

H.F. Zhang[*1], S.X.Cheng[1], L.Q. He[1], A.L. Zhang[1], Y. Zheng[1], and D.Y.Gao[1,2]

[1]Department of Thermal Science and Energy Engineering, University of Science and Technology of China,Hefei, China, 230027

[2]Department of Mechanical Engineering, University of Kentucky, Lexington, KY 40506-0108

ABSTRACT

In this paper, a new technique, using a tiny thermistor with 0.3~0.5mm in diameter to determine thermal conductivity of biomaterials in wide temperature range, has been developed. Based on steady spherical heat transfer in an infinite homogeneous medium, thermal conductivity of the measured medium can be determined by power applied and temperature rise of the thermistor. Compared with recommended values, maximum measurement errors of standard samples, aqueous glycol and $CaCl_2$ solutions, water and ice, are 5.1% in temperature range 233~313K. The thermal conductivities of rabbit's liver, kidney, heart and carotid artery in temperature range 233~293K are determined. Error caused by measurement parameters, effects of the finite scale of the measured medium and the decoupler between the thermistor and the medium are analyzed..

INTRODUCTION

With the recent advances in science and technology, heat transfer in biomedical engineering, biophysical systems and in the process of cryopreservation has drawn more and more attention [1,2]. Knowledge of thermal properties of biomaterials is becoming increasing important for diagnostic and therapeutic medicine, organ preservation and transplantation, and numerical calculation. The availability of a technique capable of characterizing accurately thermal properties of preserved organ would greatly enhance study of preservation process and therapeutic medicine. A real need exists for the accurate measurement of thermal properties in low temperature. As a basic transport property, data of thermal conductivity, well characterized with respect to tissue type, composition, and temperature, is limited. Moreover biomaterials are always soft and very easily damaged during the procedure of sample handling.

The transient hot-wire method has been successfully applied to measure the thermal conductivity of gases, liquids and solids [3-5], but it is very difficult in operation on tissues directly. A method derived from the hot-wire, using a stainless steel tube as a metallic sheath enclosed with linear heat source [6,7], can be used to measure biomaterials conveniently. Because the probe is about 30mm in length, it is still difficult to measure many tissues with small scales. Some techniques such as the 3ω method [8,9] do not require the samples to be in thermal equilibrium, which means the measurement does not require the experiment to proceed incrementally in steps of a fixed base temperature, thus making the measurement rather fast. However, the sample preparation may require substantial time, and bring more damages to the tissues because of the resistive thermometer deposition. The spherical thermistor is very small in diameter and can be used as temperature sensor and heat source simultaneously, meanwhile the damage of inserting it into the tissues is not significant, so techniques based on certain thermal model has been developed to measure local properties of tissues [10,11].

To authors' knowledge, the thermistor technique has never been used in low temperature. In this study, attempt is addressed to such issue. A tiny thermistor is used as spherical heat source inserted in an infinite measured medium, and thermal conductivity of the medium can be calculated by the temperature rise of the thermistor and the power applied. Characteristics of the technique are as following: (1) it can be used to determine thermal conductivity of organ as small as 5mm in diameter; (2) it can be used in relative wide temperature range; (3) the typical measurement time is about 30~90 seconds.

Compared with recommended values, maximum measurement errors of standard samples, aqueous glycol and $CaCl_2$ solutions, water and ice, are 2% in temperature 293K and

* Corresponding author. Email: hfzhang@mail.ustc.edu.cn

5.1% in temperature range 233~313K. A contrastive experiment is designed to validate the measurement technique for biomaterials by another method, which is developed by Cheng [7]; the difference between them is less than 4%. Additionally, error caused by measurement parameters, effect of finite scale of the measure medium and effect of the decoupler between the thermistor and the measured medium are discussed respectively.

THERMAL MODEL

In the idealized model, a spherical thermistor is inserted in an infinite homogeneous measured medium. Assuming that (1) thermistor and medium are in thermal equilibrium before heated, (2) there is no contact thermal resistance between thermistor and medium, (3) in the process of measure, the radius and thermal conductivity of the thermistor keep constant, (4) heat-source is homogeneous in thermistor, when thermistor is heated by an electric power, temperature rise of the thermistor (θ_p) and the measured medium (θ_m) in steady state can be described by the following control equations and boundary conditions

$$\nabla^2\theta_p + \frac{P}{V_p}\cdot\frac{1}{k_p} = 0, \quad r \leqslant a \tag{1a}$$

$$\nabla^2\theta_m = 0, \quad r > a \tag{1b}$$

$$\theta_p = \theta_m, \quad r = a \tag{2a}$$

$$k_p\frac{\partial\theta_p}{\partial r} = k_m\frac{\partial\theta_m}{\partial r}, \quad r = a \tag{2b}$$

$$\frac{\partial\theta_p}{\partial r} = 0, \quad r = 0 \tag{2c}$$

$$\theta_m = 0, \quad r \to \infty \tag{2d}$$

where r is the radial coordinate measured from the center of the thermistor, k_m and k_p are thermal conductivities of the measured medium and the thermistor respectively, a is the radius and V_p is the volume of thermistor, and P is the power applied. The solutions of the question above are

$$\theta_p = \frac{P}{4\pi k_m}\left[\frac{1}{2a^3}\frac{k_m}{k_p}(a^2 - r^2) + \frac{1}{a}\right] \tag{3}$$

$$\theta_m = \frac{P}{4\pi k_m}\frac{1}{r} \tag{4}$$

Because actual temperature measured is the average value in whole volume of thermistor, the temperature rise of the thermistor ($\triangle T$) is

$$\Delta T = \frac{1}{V_p}\int_0^a 4\pi r^2\theta_p dr = \frac{P}{4\pi a}\left(\frac{1}{k_m} + \frac{1}{5k_p}\right) \tag{5a}$$

Or

$$\frac{1}{k_m} = 4\pi a\frac{\Delta T}{P} - \frac{1}{5k_p} \tag{5b}$$

Obviously, if a and k_p of the thermistor is known, thermal conductivity of the measured medium can be determined from equation(5b) by measuring $\triangle T$ and P.

EXPERIMENT

Detector

A tiny spherical thermistor, 0.3~0.5mm in diameter, is fitted on one side of a stainless steel tube, 0.6mm in outer diameter, by epoxy resins, and two leads of the thermistor go through the tube, figure 1 shows the structure of the detector. In measurement experiment, the detector is inserted into the measured medium and connected to the measurement circuitry (figure 2) by the leads

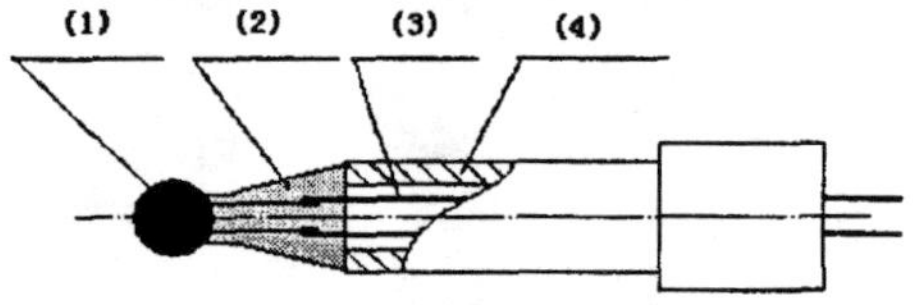

Figure 1. Structure of the detector
(1) thermistor (2) epoxy resins
(3) leads (4) stainless steel tub

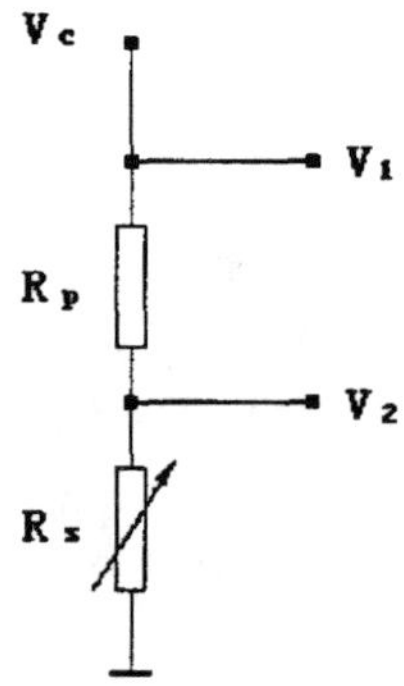

Figure 2. Measurement Circuitry
V_c—DC voltage regulator R_p—thermistor
R_S—standard resistance

Protocol

The measurement circuitry is shown in figure 2, where R_p is resistance of the thermistor, R_S is an adjustable precision resistor, and V_c is a DC voltage regulator. The thermistor is first used passively to sense the temperature of the measured medium, and then self heated. When end voltages (V_1 and V_2) reach steady state, resistance of thermistor and the power supplied can be calculated

$$R_p = (V_1 - V_2)R_S / V_2 \tag{6}$$

$$P = (V_1 - V_2)V_2 / R_S \tag{7}$$

Relationship of thermistor's resistance and the temperature can be described as following [12]

$$R_p = R_{p0} \exp[\beta(\frac{1}{T} - \frac{1}{T_0})] \quad (8)$$

where T and T_0 are temperature values corresponding to resistance of the thermistor R_p and R_{p0} respectively, and β is the coefficient of the thermistor。From equation(8), we can get the temperature rise ($\Delta T = T - T_0$) of the thermistor

$$\Delta T = \frac{TT_0}{\beta} \ln(\frac{R_{p0}}{R_p}) \quad (9a)$$

In measurement experiment, if the temperature rise ΔT is controlled less than 3K, equation(9a) can be written approximately as

$$\Delta T \approx \frac{T_0^2}{\beta} \ln(\frac{R_{p0}}{R_p}) \quad (9b)$$

In measurement experiment, after self-heated, resistance, power applied and temperature of the thermistor change with the time. From experiments of different media, the measurement time can be determined by observing the end voltage change less than 10μV per second, which means the temperature change of the thermistor is less than 0.0014K per second. Generally, the measurement is 30~90 seconds.

Calibration

β

An IceCube, a Computer Controlled Freezer with temperature control precision ±0.03K, provides the appointed temperature from 223 to 313K. Resistance of the thermistor in different temperature is recorded. Coefficient β can be calculated by least squares method from equation (8). Figure 3 shows the calibration result, where the scattered points are the measured resistance values, and the continuous curve is fitted result with β= 3414K.

a and k_p

Table 1. Calibration result of a and k_p

Standard Sample	Glycerin	Agar-gelled Water
Conductivity (W/mK)	0.286	0.596
$\Delta T/P$ (K/W)	1467	873.7
$\Delta T/P$ Reproducibility (%)	0.4	0.8
a (mm)	0.244	
k_p (W/mK)	0.208	

Glycerol and agar-gelled water (which contains 5% agar in order to prevent convection, and thermal conductivity of water is not significantly altered by the presence of agar) [11,13,14] are chose as the standard samples to calibrate a and k_p of the thermister in 293K. In process of calibration, through regulating the supplied power from 2 to 5 mill-Watt, temperature rise of the thermistor can be controlled less than 3K. Twelve experiments are conducted for both liquids and the measured ΔT and P are recorded, after discarding the highest and lowest readings, the average $\Delta T/P$ are calculated. Table 1 shows the calibration result, a=0.244 mm and k_p=0.208 W/mK.

Figure 3. Experimental result of thermistor's resistance and temperature

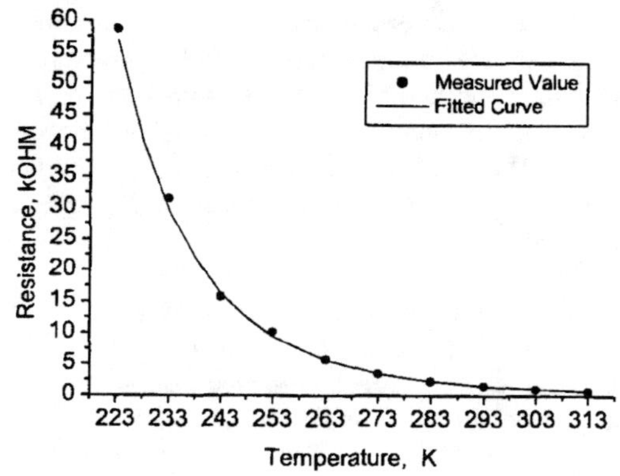

RESULTS

In temperature 293K

Based on the known β, a and k_p, thermal conductivity of the measured medium can be determined by measuring $\Delta T/P$ of the thermistor. In order to assess reliability and precision of the technique, some standard samples with recommended values [13,14] are measured. The experimental results are listed in table 2, and the maximum relative error is 2%. Because of the convection of liquid, the measured value is always bigger than the real value in table 2.

Table 2. Measurement results of some standard samples in 293K (W/mK)

Sample	Measured Value	Recommended Value[14]	Error(%)
Glycol	0.258	0.256	0.8
Glycol Solution 23.4%wt	0.497	0.488	1.8
Glycol Solution 46.4%wt	0.436	0.430	1.4
$CaCl_2$ Solution 29.9%wt	0.559	0.548	2.0

Table 3. Measurement results of some biomaterials in 293K (W/mK)

Sample	Measured Value (our method)	Measured Value (probe [14])	Error (%)
Thawed Muscle	0.464	0.480	3.3
Thawed Fat	0.217	0.223	2.7
Apple	0.377	0.370	1.9

For biomaterials, because their thermal conductivities are different due to different composition and tissue type, there is no proper recommended value presently. Therefore, the probe [7] is set up to measure the same sample, which has a relative large scale, with our detector simultaneously in order to compare the experimental results. The contrastive results are listed in table 3, and the difference between the two methods is less than 4%.

In temperature range 233～313K

The thermistor is mainly composed of porcelain, and some kinds of metals [12], and their properties of thermistor are very stable in the experimental temperature range [16], so it is a reasonable assumption that the radius and thermal conductivity of thermistor in different temperature is same as that in temperature 293K. This makes the experiments in different temperatures simplified greatly, and the measurement results prove the assumption.

Table 4 and 5 list the results of aqueous glycol 46.4%wt solution measured from 243 to 313K, and pure water or ice from 233 to 293K. Compared with the recommended values [14,15] the maximum relative errors are 2.6% and 5.1% respectively.

Table 4. Measurement results of glycol 46.4%wt in temperature 243~313K (W/mK)

Temperature (K)	313	303	293	283	273	263	253	243
Measured Value	0.438	0.436	0.436	0.431	0.430	0.441	0.437	0.439
Recommended Value [14]	0.430	0.430	0.430	0.430	0.430	0.430	0.430	0.430
Error (%)	1.9	1.4	1.4	0.2	0	2.6	1.6	2.1

Table 5. Measurement results of water and ice in temperature 233~263K (W/mK)

Temperature (K)	313	303	293	283	273	263	253	243
Measured Value	0.438	0.436	0.436	0.431	0.430	0.441	0.437	0.439
Recommended Value [14,15]	0.430	0.430	0.430	0.430	0.430	0.430	0.430	0.430
Error (%)	1.9	1.4	1.4	0.2	0	2.6	1.6	2.1

Experimental results in table 2.~5. have demonstrated reliability and precision of the technique. After that, some fresh organs of rabbit in vitro approximately an hour post mortem are determined from 233 to 293K. Table 6 lists the data of experimental result and figure 4 shows the change of thermal conductivities of the determined organs with temperature, compared with that of water or ice.

Table 6. Measurement results of some fresh rabbit's organs in temperature 233~293K (W/mK)

Temperature (K)	293	283	273	263	253	243	233
Liver	0.522	0.509	0.502	1.04	1.39	1.57	1.75
Kidney	0.493	0.485	0.480	0.895	1.17	1.33	1.49
Heart	0.512	0.498	0.490	1.01	1.22	1.48	1.70
Carotid Artery	0.414	0.397	0.388	0.902	1.23	1.42	1.56

During the experiments, organs are enclosed in a sealed container, because the air current in the IceCube makes the organs dehydrate quickly. For the carotid artery, filter paper is used to extract liquid from their surface, and then they are put with inserted detector into the sealed container, which is 10mm in diameter. From table 6 and figure 4, it can be found that (1) the difference of thermal conductivities of any organ in different temperature is significant; (2) below temperature 273K, thermal conductivities of organs increase significantly as temperature decreases; (3) change of organs' thermal conductivities is like that of the water or ice, but their values are different significantly.

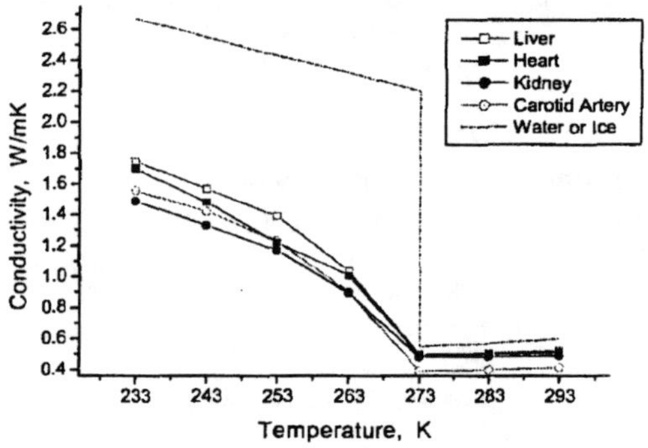

Figure 4. Thermal conductivities of rabbit's organs from 233 to 293K

ERROR ANALYSIS

Error caused by the measurement parameters

From equation (5), the relative error of measured thermal conductivity can be described as:

$$\left|\frac{\delta k_m}{k_m}\right| \le 4\pi a\frac{\Delta T}{P}k_m\cdot[\left|\frac{\delta a}{a}\right|+\left|\frac{\delta P}{P}\right|+\left|\frac{\delta(\Delta T)}{\Delta T}\right|]+\frac{k_m}{5k_p}\left|\frac{\delta k_p}{k_p}\right| \qquad (10)$$

From typical measurement parameters coefficient $4ak_m\Delta T/P \approx 1.5$. The error of applied power P that is determined by voltage measurement is very small, so its relative error can be negligible. Assuming the drift of the base temperature in the experiments causes the error of ΔT, the relative error of ΔT is about 1% by using the temperature control precision of the instrument.

Error of a and k_p is mainly caused by the calibration, from table1 the calculated results are:

$$\left|\frac{\delta a}{a}\right| \le 0.6\% \qquad \left|\frac{\delta k_p}{k_p}\right| \le 2.5\%$$

AS the measured k_m increases from 0.4 to 2.0W/mK, the error caused by k_p reaches 5%.

Summarily, the total measurement error<7.5%, and the main error comes from the thermal conductivity of the thermistor.

Effect of the decoupler between the thermistor and the measured medium

An imperfect thermal contact could result from a thin air or liquid film between the thermistor and the medium. We estimate its effect by the same way as that in previous paragraph. Calculation is run with typical parameters and changes of the decoupler thickness form 0.01 to 0.05 mm, and thermal conductivity of the decoupler from 0.02 to 2.0 W/mK independently. Result shows that the temperature change of the thermistor is less than 1%, as the thermal conductivity difference between the medium and the decoupler is less than 50%. If the medium is biomaterial, the error will decrease further, because the decoupler is normally filled with the tissue liquid and thermal conductivities of the tissue and tissue liquid are almost the same. Therefore, thermal contact resistance is negligible during measuring biomaterials. Validity of the analysis above can be proved by the experimental result in table 3.

CONCLUSION

The technique developed in this study is convenient and reliable to determine thermal conductivities of liquid and biomaterials in temperature 233~313K, and especially suit to small samples.

The lowest temperature in this study is 233K. The reason of this restrict is that the characteristics of the thermistor. From fig3, the resistance of the thermistor increases rapidly as the temperature decrease. When temperature reaches 223K, it is about 60kΩ, which will make the thermistor unsteady. Although the experimental temperature range is not enough to study the full process of cryopreservation, it shows that thermal conductivity of organs below zero is significantly different from that of ice. If a thermistor with a smaller coefficient β is available, thermal conductivity could be determined by the proposed technique.

ACKNOWLEDGMENTS

The authors acknowledge the Project of Hundreds of Outstanding Scholars of the Academy of China(2000-2003), NSF of China (No.59976041, 50106016), and of Anhui Province (No.00047520) for support of this work.

REFERENCES

1. Chato J C and Lee R, 1998, The future of biothermal engineering Ann., NY Acad. Sci. Vol.858, pp. 1-20
2. Diller K R and Ryan T P, 1998, Heat transfer in living systems: current opportunities, J. Heat Transfer Vol.120, pp.810-29
3. Blackwell J H, 1954, A transient flow method for determination of thermal constants of insulating materials in bulk., Part I—theory., J. Appl. Phys., Vol.25, pp.137-44
4. Andersson P and Backstrom G, 1976, Thermal conductivity of solids under pressure by transient hot-wire method, Rev. Sci .Instrum., Vol.47, pp.205-09
5. Banaszkiewidz M, Seiferlin K, Spohn, et al., 1997, A new method for the determination of thermal conductivity and thermal diffusivity from linear hear source measurements, Rev. Sci. Instrum., Vol.68, pp.4184-90.
6. Liang X G, Ge X S, Zhang Y P and G J Wang, 1991, A convenient method of measuring the thermal conductivity of biological tissue, Phys. Med. Biol., Vol.36, No 12, 1599-05
7. Cheng S X, Jiang Y F, Liang X G, 1994, A Tiny Heat Probe for Measuring the Thermal Conductivities of Non-rigid Materials, Meas. Sci. Technol., Vol.5, pp.1339-44.
8. Gahill D G, 1990, Thermal conductivity measurement from 30 to 750K: 3ω method. Rev. Sci. Instrum. Vol.61, pp.802-08
9. Lu L, Yi W and Zhang D L, 2001, 3ω method for specific heat and thermal conductivity measurements, Rev.Sci.instrum. Vol.72, pp.2996-03
10. Arkin H, Holmes K R, Chen M M and Bottje W G, 1986, Thermal Pulse Decay Method for Simultaneous Measurement of Local Thermal Conductivity and Blood Perfusion: A Theoretical Analysis, ASME Journal of Biomechanical Eng., Vol.108, pp.208-14.
11. Vovalo J W, Allen J T, and Bowman H F 1984 The Simultaneous Measurement of Thermal Conductivity, Thermal Diffusivity and Perfusion in small Volumes of Tissue, ASME Journal of Biomechanical Eng., Vol.106, pp.192-97.
12. Liu D F, 1993, Technique and Application of Semi-conductive Sensor (Dalian: DLUT Press) in China
13. Toulokian Y S, Liley P F and Sanena S C, 1970, Thermophsical Properies of Material, Vol.3 (New York:IFI/plenum)
14. Chen Z S, Ge X S and Gu Y Q, 1990, Measurement of Heat and Thermophysical Properties (Hefei: USTC Press) in China
15. Fukusako S, 1990, Thermophysical Properties of Ice, Snow, and Sea Ice, International Journal of Thermophysics, Vol.11, No.2.
16. Chen G B, 1998, Cryogenic Engineering Materials (Hongzhou: ZJU Press) in China.

Proceedings of IMECE2002
ASME International Mechanical Engineering Congress & Exposition
November 17-22, 2002, New Orleans, Louisiana

IMECE2002-33688

MEASUREMENT OF CRYOPROTECTANT CONCENTRATION USING MRI IN PSEUDOBIOLOGICAL TISSUES AND DETERMINATION OF DIFFSIVITY BASED ON INVERSE PROBLEM ANALYSIS

Hiroshi ISHIGURO*
Department of Biological Functions and Engineering, Graduate School of Life Science and Systems Engineering,
Kyushu Institute of Technology,
1-1 Sensui-cho, Tobata-ku, Kitakyushu, Fukuoka 804-8550, JAPAN

Takanori KAI
Graduate School of Engineering Mechanics, University of Tsukuba, Tsukuba, Ibaraki 305-8573, JAPAN

Kuniyasu OGAWA
Department of Mechanical Engineering, Keio University, Yokohama, Kanagawa 223-8522, JAPAN

ABSTRACT

Transient one-dimensional distribution of cryoprotectant concentration in pseudobiological tissues (agar) was measured noninvasively using magnetic resonance imaging (MRI). Cryoprotectants were dimethyl sulfoxide (DMSO) and glycerol, common cryoprotectants penetrating cells. Attenuation of MRI image intensity due to volumetric fraction of solution and relaxation times was also investigated. Apparent diffusivity of each cryoprotectant as a function of agar concentration was determined from the inverse problem analysis. The diffusivity decreased with an increase in agar concentration. This method was also applied to the liver tissues of chicken.

NOMENCLATURE

C : cryoprotectant concentration of solution in medium
D_S : apparent diffusivity of cryoprotectant
DMSO : dimethyl sulfoxide
H : magnetic field
I : MRI image intensity
$\tilde{I}$: nondimensional intensity of MRI image
m : agar concentration (mass of agar/mass of solvent)
MRI : magnetic resonance imaging
MW : molecular weight
NMR : nuclear magnetic resonance
SG : specific gravity
t : time from start of measurement
Te : echo time
Tr : repetition time
T_1 : longitudinal relaxation time
T_2 : transverse relaxation time
x: distance in axial direction of test tube from boundary between mass transfer medium and solution
y : distance in radial direction of test tube
ε : volumetric fraction of solution
ρ : nuclear magnetic moment density of proton bonded with particular kind of atom

INTRODUCTION

Cryopreservation, preservation by freezing, is available to preserve biological materials for transplantation: biological cells and tissues, and bioartificial tissues. Cryoprotectants, chemical compounds to improve viability of the biological materials through freezing and thawing, need to be introduced into the materials before freezing and to be removed from the materials after thawing. Cryopreservation consists of five processes: 1)introduction process of cryoprotectants into the materials, 2)freezing process, 3)preservation process, 4)thawing process, and 5)removal process of cryoprotectants from the materials. Large step change in external osmolarity due to high concentration of cryoprotectants in the process 1) and/or low concentration of cryoprotectants in the process 5) can reduce the cellular viability. The former induces excessive dehydration and shrinkage of cells and consequently can cause damage to the materials. The latter induces excessive inflow of water into cells and excessive expansion of cells, consequently causing damage. Long-term exposure of materials to even low concentration of cryoprotectants in the processes 1) and 5) manifests toxicity of the cryoprotectants and can be also lethal. Therefore, optimization of cryoprotectant protocols in relation to the above two processes 1) and 5) is important for success of cryopreservation. Cryopreservation benefits from measurement of

*Corresponding author's e-mail: ishiguro@life.kyutech.ac.jp

cryoprotectant concentration in the biological materials and investigation of permeation rate of cryoprotectants into the materials.

Permeation of cryoprotectants into cells and tissues is accompanied with permeation of water. For a single cell, time-variation of cell volume in hypertonic solution was measured with a light microscope, and the membrane permeation parameters of water and cryoprotectant were estimated by inverse problem analysis (Schwartz and Diller, 1983; McGrath, 1997; Diller et al., 1988). The methodology for a single cell is almost established.

For tissues and organs, the measurement of permeation of water and cryoprotectants and the estimation of permeation rate of cryoprotectants is not so easy as those for a single cell because the tissues and organs consist of a lot of cells and have complicated structures. Direct measurement of cryoprotectant in the tissues was challenged. For example, during the permeation of DMSO into a common carotid artery of a rabbit, the amount (overall concentration) of DMSO in the tissues was measured by means of NMR spectroscopy (Bateso et al., 1994), where any information of spatial distribution was lost. Fast measurement by MRI of DMSO concentration distribution in an artificial engineered dermal replacement was also tried (Bidault, et al., 2000). For organ tissues, DMSO concentration was measured by MRI during the permeation into the slices of rat liver and rat kidney to estimate effective permeation rates as a function of temperature on trial (Isbell et al., 1997). For a whole organ, permeation of glycerol in the organ with perfusion of solution was estimated by using the network thermodynamics model to investigate influence of perfusion protocols on the permeation (Lachenbruch et al., 1998). The methodology for measurement of permeation of cryoprotectants into the tissues and organs remains to be established.

Primarily, transient spatial distribution of cryoprotectant concentration need to be measured precisely to investigate the permeation of cryoprotectant. In this study, magnetic resonance imaging (MRI), a noninvasive method applicable to biological materials, was applied to the measurement of cryoprotectant concentration during the diffusion in pseudobiological tissues (agar) and solutions. Preliminarily, proton-nuclear magnetic resonance (^{1}H-NMR) spectrum, chemical shift frequency, and relaxation times were measured in NMR experiments. Calibration curve between MRI image intensity and cryoprotectant concentration was measured for the pseudobiological tissues and solution by means of chemical shift imaging (CSI) technique. Attenuation of MRI image intensity for the pseudobiological tissues compared with that for the solution was considered on volumetric fraction of solution and relaxation time. Transient one-dimensional permeation of DMSO and glycerol, common cryoprotectants penetrating cells, was imaged. Apparent diffusivity of each cryoprotectant as a function of agar concentration was determined by means of the inverse problem analysis based on the time-series one-dimensional concentration-distribution of cryoprotectant. This MRI method was also applied similarly to fresh liver tissues of chicken.

EXPERIMENTAL MATERIALS

Agar (Quick soluble agar UP-37, Ina Food Corporation, Japan) was selected as pseudobiological tissues. Agar concentration was set at four values: m=0.004, 0.016, 0.040, and 0.067. Two common cryoprotectants penetrating cells were used: DMSO ($(CH_3)_2SO$, MW=78.12, SG=1.101) and glycerol ($C_3H_5(OH)_3$, MW=92.10, SG=1.261). Molecules of these cryoprotectants have protons bonded with carbon molecules unlike water. Fresh liver taken from a 40- to 45-day-old broiler (by Arbor Acres) immediately after slaughter was also used for the tissues with homogeneous structure.

EXPERIMENTAL APPARATUS AND METHODS

^{1}H-NMR spectrum, chemical shift frequency, and relaxation times for the protons bonded with particular kind of atoms were measured by means of NMR for each sample: distilled water, pseudobiological tissues, and tissues with various concentration of cryoprotectant. Molecules of water and cryoprotectant were imaged separately by means of CSI technique, and transient spatial distribution of cryoprotectant concentration was measured. The whole measurement was performed at temperature of around 20°C.

Magnetic Resonance Imaging System

MRI system used in this study was a Unity Inova 300/150 SWB produced by Varian. The specification was 7.05T in magnetic field, 300MHz in resonance frequency, 20G/cm in maximum gradient magnetic field, and 150mm in inner diameter of magnet.

Test Section and Methods

Figure 1 shows the test section in the probe for detection of NMR signal. Two glass test tubes were arranged vertically in the central part of the probe with 25mm inner diameter.

In measurement by NMR, one of the two test tubes was filled with hexamethyldisiloxane (HMDS, $[Si(CH_3)_3]_2O$), MW=162.38) which worked as reference in measuring chemical shift. The other was filled with each sample (solution, agar, tissues).

In measurement of MRI, one of the two tubes was filled with 5.0M cryoprotectant aqueous solution, and it was used to compensate for spatial-temporal inhomogeneity of MRI image intensity. The other was for measurement of 1)calibration curve between cryoprotectant concentration and MRI image intensity and 2)transient spatial concentration-distribution during the diffusion of cryoprotectant. For 1), the test tube was full of cryoprotectant

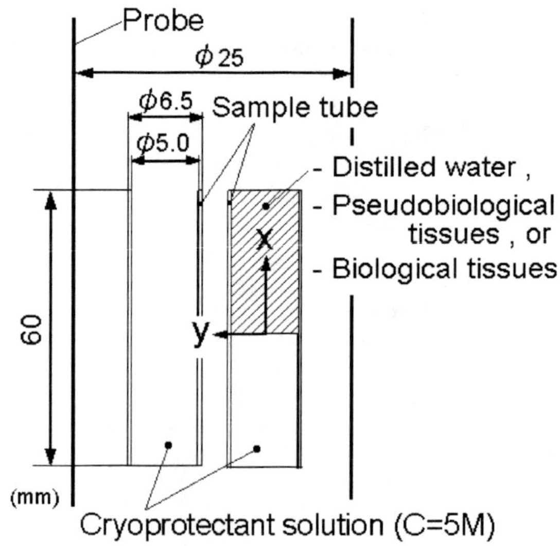

Figure 1 Test section

solution of predetermined concentration (C=0, 1.0, 2.0, 3.0, 4.0, 5.0, 6.0M) or agar prepared with each solution. For 2), distilled water or agar prepared with distilled water was put on 5.0M cryoprotectant solution to stabilize the fluid stratified with density, consequently forming unsteady one-dimensional diffusion.

Measurement by Nuclear Magnetic Resonance

For each sample, ^{1}H-NMR spectrum, longitudinal relaxation time, and transverse relaxation time, were respectively measured by the pulse Fourier transformation (PTF) method, inversion-recovery (IR) method, and Carr-Purcell-Meiboom-Gill (CPMG) method (Callaghan, 1991).

Chemical Shift Imaging

Molecules of cryoprotectant in the solution included in the mediums can be imaged separately from water molecules to measure the concentration of cryoprotectant proportional to the nuclear magnetic moment density of protons bonded with carbon by using CSI technique (Callaghan, 1991). Conversely, just water molecules can be also imaged. The technique is based on the principle that protons bonded with carbon atom of CH_3, CH_2, and CH and proton bonded with oxygen atom of OH have an individual different resonance frequency.

One frame of image obtained by MRI was 128×64 pixels (51.2mm×20.0mm, 1.0mm in slice thickness). Te was set at 0.018sec for all the samples. Copper sulfate $CuSO_4$ was used as a contrast medium on trial to shorten relaxation time, repetition time, and acquisition time per frame of image. For the solution and pseudobiological tissues with $CuSO_4$, Tr was set at 3.0sec (1.0sec only for glycerol solution), acquisition time per frame was 384sec (128sec only for glycerol solution), and acquisition period was 15min. For the solution and pseudobiological tissues without $CuSO_4$, Tr was set at 15sec, acquisition time per frame was 16min, and acquisition period was 30min. Cryoprotectant concentration was measured for 6hr during the diffusion. Finally, the use of $CuSO_4$ was avoided considering to apply to the biological materials. It was confirmed that the measurement without $CuSO_4$ did not cause different results of concentration distribution and diffusivity from that with $CuSO_4$ because of very slow diffusion of cryoprotectant.

EXPERIMENTAL RESULTS AND DISCUSSION

^{1}H-NMR Spectrum and Chemical Shift Frequency

In the agar without cryoprotectant, no signal from the protons bonded with carbon atoms was detected, and only signal from the proton of water was detected. By contrast, in the agar with cryoprotectant the former signal was detected in distinction from the latter signal. Chemical shift frequency of the proton of cryoprotectant from that of H_2O was 613Hz in CH_3 of DMSO, and 391Hz and 338Hz in CH_2 and CH of glycerol, respectively. The signal from particular proton of cryoprotectant was able to be detected selectively on the basis of this chemical shift.

Relaxation Times

MRI image intensity $I \propto \rho(H)\cdot[1-\exp(-Tr/T_1)]\cdot\exp(-Te/T_2)$, which means that I is dependent on T_1 and T_2 as well as ρ. Relaxation times of protons bonded with carbons of cryoprotectant for the agar decreased with an increase in agar concentration compared with the values for the solution, and the relaxation times also decreased with an increase in cryoprotectant concentration. The image intensity was not influenced by the decrease in T_1, but the intensity decreased monotonously with a decrease in T_2.

Volumetric Fraction of Solution in Pseudobiological Tissues

Solid components included in the pseudobiological tissues and biological tissues decreased nuclear magnetic moment density ρ because of the decrease in volumetric fraction of the solution in the medium compared with only the solution, and consequently attenuated the image intensity of MRI. The volumetric fraction of solution ε was defined as the ratio of image intensity, compensated for influence of relaxation time, from the OH-excitation for the agar to that for only the solution. For example, ε had a value of 0.93 for the agar with agar concentration of 0.067. The above defined ε was very consistent with the volumetric fraction of solution estimated from the increase in volume of the medium in dissolving agar-powder.

Calibration Curve between MRI Image Intensity and Cryoprotectant Concentration

Calibration curves between MRI image intensity and cryoprotectant concentration were obtained from direct measurement for the agar and solution. Figure 2 shows the typical calibration curves of the image intensity I(C) against DMSO concentration C of the solution included in the agar (m=0.040). Nondimensional intensity of MRI image, $\tilde{I}(C)$, was defined and normalized as $[I(C) - I_b]/[I_0(5) - I_b]$, where I(C) denotes the average over the x-y plane of two-dimensional image intensity from the CH_3-excitation for the particular medium in the test tube, I_b image intensity of the background, and $I_0(5)$ image intensity for 5.0M DMSO solution. Nondimensional intensity of calibration curve for the agar and that for the solution are, respectively, denoted by close triangle and open circle in Figure 2. They increased nearly in proportion to DMSO concentration, and the former was smaller than the latter by about 15%. Practically, we used the directly measured value as the calibration curve for the agar.

The volumetric fraction of solution ε and the transverse relaxation time T_2 of DMSO in the agar could attenuate the image

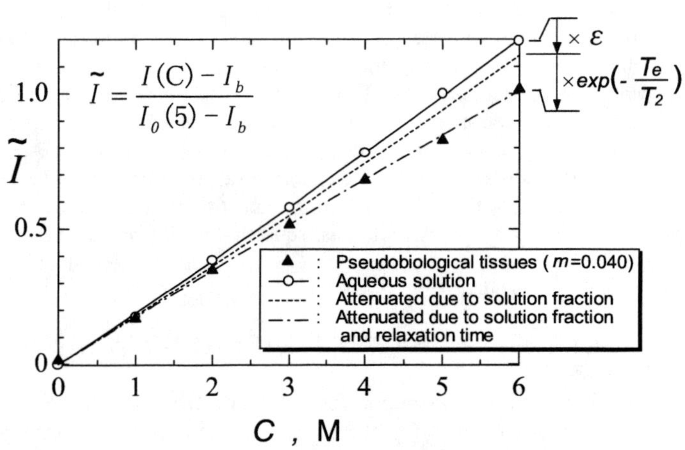

Figure 2 Calibration curves of MRI image intensity against DMSO concentration in pseudobiological tissues and aqueous solution

intensity compared with that for the solution. The values of ε and T_2 were, respectively, obtained in the previous chapter of this article. The nondimensional intensity estimated in consideration of the attenuation due to ε from the value for the solution is also shown with a broken line in Figure 2. That estimated by considering the attenuation due to T_2 in addition to ε is shown with a one-dotted chain line. The value denoted by the latter line was very consistent with the directly measured value for the agar. The result gives clear breakdown of the attenuated intensity, and indicates that the calibration curve for the agar can be estimated from that for the solution provided that ε and T_2 are measured independently. This has a practically important meaning for biological tissues, in which there is an individual difference and the repetitional use of the same sample is very difficult or impossible.

Unsteady One-Dimensional Diffusion of Cryoprotectant

Transient two-dimensional distribution of DMSO in the agar (m=0.040) is shown in Figure 3 as a typical result from the concentration measurement during the diffusion of cryoprotectant. For 6 hr, the concentration distribution was homogeneous in the y direction and gradient in only the x direction, which indicates one-dimensional diffusion. Time-series concentration distribution in the x direction, averaged in the y direction to improve the S/N ratio, is shown in Figure 4. DMSO diffused into the agar due to mutual diffusion between DMSO and water in the solution included in the agar, and DMSO concentration increased there.

INVERSE PROBLEM ANALYSIS FOR APPARENT DIFFUSIVITY

Apparent diffusivity D_S of cryoprotectant in the medium was defined according to the diffusion equation: $\partial C/\partial t=\partial(D_S\cdot\partial C/\partial x)/\partial x$, where $D_S = \alpha\cdot C + \beta$ (α and β are constants). The dependency of D_S on C was considered with a linear function because the range of concentration in this study was narrow. The diffusivity was determined by inverse problem analysis based on transient one-dimensional concentration distribution of cryoprotectant in the agar.

As boundary conditions for the analysis, time-variation of cryoprotectant concentration was given near the agar-solution boundary in the agar from the experimental result, and the mass flux of zero was imposed at the other end in the agar. (For the aqueous solution, mass flux of zero was imposed at both the ends.) As an initial condition, the concentration distribution measured at t=0 hr was given. For assumed values of α and β, transient concentration distribution was calculated from the diffusion equation, and total of squared errors between the calculated values and the experimental values was calculated to compared both the values. This procedure was repeated for such another values of two constants as to decrease the total of squared errors, and finally α and β were determined so as to minimize the total squared errors. Consequently, D_S for the agar was not dependent on C while that for the aqueous solution was dependent.

RESULTS OF APPARENT DIFFUSIVITY AND DISCUSSION

Time-series one-dimensional distribution of DMSO concentration calculated for the D_S determined in the inverse problem analysis for the agar (m=0.040) is shown with solid lines and broken lines in Figure 4 as an example and is compared with experimental values. The left end of the line corresponds to the position of boundary condition. Agreement between the both values is very good.

A series of apparent diffusivities obtained similarly for various conditions are plotted in Figure 5. The diffusivities of DMSO and glycerol decreased monotonously from the values for aqueous solution (the values averaged in the range of C=0M to 2.0M) with an increase in agar concentration. The value for DMSO was 1.5 to 1.8 times larger than that of glycerol. Also, the apparent diffusivity of DMSO in the liver tissues obtained similarly in our study is plotted with open triangle in Figure 5. The value was about 5% of the diffusivity of DMSO in aqueous solution. The liver tissues are compartmented into micro-scale pieces with cell membranes and intracellular substance, which work as resistance of mass transfer. This is the reason why the diffusivity of cryoprotectant in the

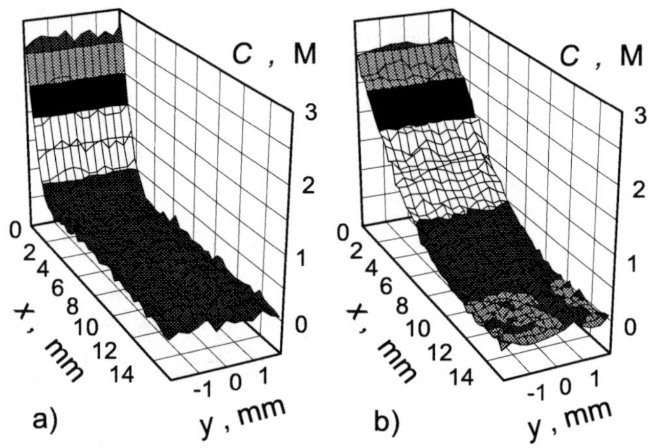

Figure 3 Transient two-dimensional distribution of DMSO concentration in pseudobiological tissues a) t=0 hr, b) t=4 hr

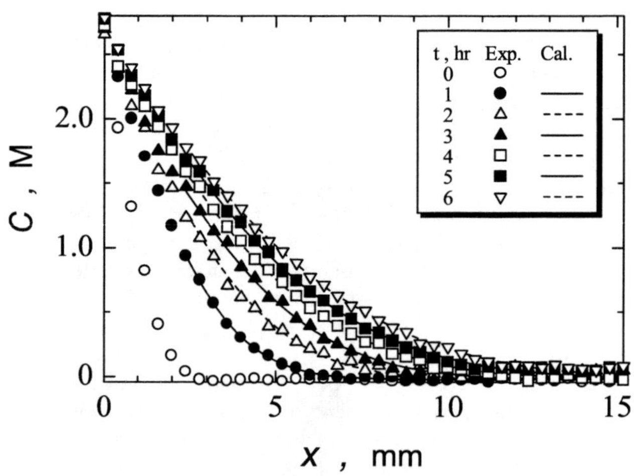

Figure 4 Transient one-dimensional distribution of DMSO concentration in pseudobiological tissues

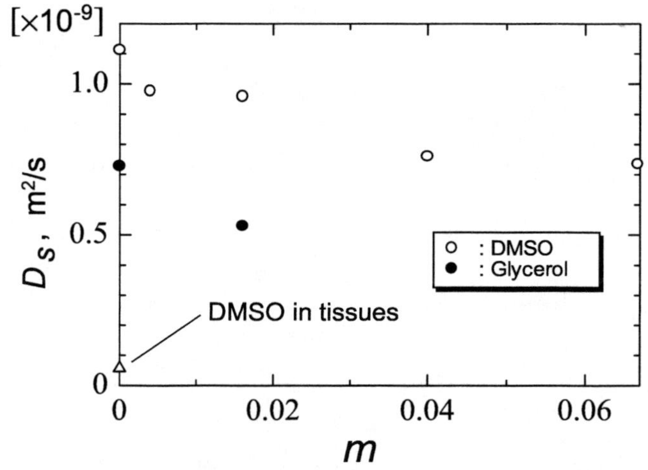

Figure 5 Apparent diffusivity of cryoprotectant in pseudobiological tissues as a function of agar concentration

biological tissues has a very small value.

In principle, the method of measuring cryoprotectant concentration using MRI can be applied to more complicated systems; 1)heterogeneous medium with ε and T_2 of spatial dependence in the medium and 2)mixture of different cryoprotectants. More research should be conducted including investigation of measurement accuracy for the complicated systems.

CONCLUSIONS

1) Cryoprotectant molecules in the pseudobiological tissues can be imaged selectively by CSI technique to measure the concentration. Transient spatial concentration-distribution of cryoprotectant can be measured during the diffusion of cryoprotectant in the pseudobiological tissues and solution.
2) Apparent diffusivity of cryoprotectant in the pseudobiological tissues as a function of agar concentration was estimated by means of inverse problem analysis based on the results from measurement of transient one-dimensional concentration distribution.
3) Calibration curve for the pseudobiological tissues can be estimated from that for the solution considering the attenuation of image intensity due to the volumetric fraction of solution and the relaxation times.

ACKNOWLEDGMENTS

The authors express their thanks for using the MRI system owned by Research Center for Carbon Recycling and Utilization, Tokyo Institute of Technology. This work was supported by Grant-in-Aid for Scientific Research (C) (#11650205) to Professor Hiroshi ISHIGURO from the Ministry of Education, Science, Sports and Culture of Japan.

REFERENCES

Bateson, E.A., Busza, A.L., Pegg, D.E., and Taylor, M.J., Permeation of Rabbit Common Carotid Arteries with Dimethyl Sulfoxide, *Cryobiology* Vol.31, pp393-397 (1994)

Bidault, N.P., Hammer, B.E., and Hubel, A., Rapid MR Imaging o Cryoprotectant Permeation in an Engineered Dermal Replacement, *Cryobiology*, Vol.40, pp13-26 (2000)

Callaghan, P.T., *Principles of Nuclear Magnetic Resonance Microscopy*, Clarendon Press, Oxford (1991)

Diller, K.R., Beaman, J.J., Montoya, J.P., and Breedfeld, P.C., Network Thermodynamic Modeling With Bond Graphs for Membrane Transport During Cell Freezing Procedures, *Journal of Heat Transfer*, Vol.110, pp938-945 (1988)

Isbell, S.A., Fyfe, C.A., Ammons, R.L.M., and Pearson, B., Measurement of Cryoprotective Solvent Penetration into Intact Organ Tissues Using High Field NMR Microimaging, *Cryobiology*, Vol.35, pp165-172 (1997)

Lachenbruch, C.A., Diller, K.R., and Pegg, D.E., Sensivity of Kidney Perfusion Protocol Design to Physical and Physiological Parameters, *Biotransport: Heat and Mass Transfer in Living Systems*, Annals of the New York Academy of Science, Vol.858, pp298-309 (1998)

McGrath, J.J., Quantitative Measurement of Cell Membrane Transport: Technology and Applications, *Cryobiology* Vol.34, pp315-334 (1997)

Schwartz, G.J., and Diller, K. R., Osmotic Response of Individual Cells During Freezing, I. Experimental Volume Measurements, *Cryobiology*, Vol.20, pp61-77 (1983)

Proceedings of IMECE2002
ASME International Mechanical Engineering Congress & Exposition
November 17–22, 2002, New Orleans, Louisiana

IMECE2002-33693

MULTI-PARAMETER ESTIMATION IN HYPERTHERMIA PROBLEM BY USING AN OPTIMAL CHOICE OF DESCENT PARAMETERS IN ITERATIVE METHODS

Tahar Loulou
Ecole des Mines d'Albi-Carmaux
81000, Albi,
France
E-mail: loulou@enstimac.fr

Elaine P. Scott*
Mechanical Engineering Department
Virginia Tech, Blacksburg, VA 24061-0238
U.S.A.
E-mail: scottep@vt.edu

Abstract

This paper describes a numerical procedure conducted to estimate thermo-physical properties of the human tissue during hyperthermia treatment of a cancerous region. The estimation algorithm is based on the solution of an inverse heat conduction problem. The Gauss-Newton method is used to estimate simultaneously the volumetric heat capacity, the thermal conductivity, and the volumetric blood rate (blood perfusion) in the bio-heat transfer equation during a hyperthermia treatment cycle. The treatment quality of hyperthermia is analyzed by the computation of the thermal dose which is obtained from the resulting temperature field in the tissue. The importance of an accurate estimation of the thermo-physical properties of the tissue lies in that they are the most important factors for achieving a high precision heating cycle which results in an optimized treatment. The inverse analysis is based on the temperature measurements taken inside the cancerous tissue region during the transient heating process. An experimental optimization procedure is conducted to make the estimated parameters as accurate as possible. Several numerical tests were performed and show that the developed method provides an accurate estimation of thermo-physical properties in a very short practical time. As the blood perfusion is very sensitive to the temperature variation in the tissue, this estimation tool can be implemented during one cycle treatment which results in an on-line thermo-physical parameter correction while the treatment is performed.

*Professor, Corresponding author.

Nomenclature

c	volumetric specific heat of tissue,
k	tissue thermal conductivity,
L	length,
N_p	number of unknown parameter,
N_s	number of sensors,
N_t	number of measurements,
$Q(x,t)$	heating power,
$S(U)$	residual function,
$T(x,t)$	temperature,
T_b	arterial blood temperature,
U	vector of unknown parameters,
u_i	i^{th} unknown parameter,
t, t_f	time, final time,
w	volumetric blood diffusion rate,
x, x_i	space, sensor location,
X_i	sensitivity coefficient,
$Y(x,t)$	measured temperature,
ε_i	computation error for the ith parameter

Introduction

Often High Intensity Focused Ultrasound treatments (HIFU) scanned or phased array ultrasound are used to heat a specified tissue (e.g., tumor) region. Multiple pulses are applied for short periods of time, and tissue necrosis follows an Arrhenius relationship between temperature and time. In optimizing the thermal dose or time-temperature treatment, the applied thermal dose is compared to the prescribed one with the goal of assuring an optimal thermal dose in the specified tissue (e.g. tumor) region, and a minimal thermal dose in the normal tissue region. The computation of a thermal dose is obtained from the resulting temperature field in the tissue and depends strongly on the used thermo-physical properties.

Thus, the precise knowledge of thermo-physical parameters characterizing the thermal model is required and indispensable to achieve a high precise treatment. It is well known that in the solution of inverse problems it often becomes necessary to determine several independent functions or parameters at one time. This is why parameter estimation tools are needed to estimate these unknown coefficients and to investigate their influence on the heating process quality. Its importance is well documented, detailed and explained by Beck [1] in his excellent book : *Parameter estimation provides tools for the efficient use of data in the estimation of constants appearing in the mathematical models and for aiding in modelling of phenomena.*

Based on the temperature measurements inside the tissue, the presented estimation procedure uses the Gauss-Newton [1, 2] method to estimate simultaneously three coefficients used in the bio-heat model: specific heat, thermal conductivity and the volumetric blood diffusion rate of the tissue by minimizing an objective function. This technique has been widely used in parameter estimation in different engineering fields and sciences.

This paper is divided in six major sections. The mathematical modelling of unsteady bio-heat transfer in a direct problem form is shown in section two. The formulation of the parameter estimation problem as an inverse heat coefficient problem and the Gauss-Newton method are presented in section three. Numerical results with an elementary sensitivity analysis are given in section five. The last section presents some concluding remarks.

Direct problem

Heat transfer in bio-mechanical systems is often modelled by using Pennes Bio-heat model [3]. The one dimensional heat transfer equation for one layer of thickness L is described the following system of equations:

$$c\,\frac{\partial T(x,t)}{\partial t} = k\,\frac{\partial^2 T(x,t)}{\partial x^2} + w\,[T_b - T(x,t)] + Q(x,t) \qquad (1)$$

$$T(x,t) = T_b, \qquad x = 0, \qquad 0 < t \leq t_f \qquad (2)$$

$$T(x,t) = T_b, \qquad x = L, \qquad 0 < t \leq t_f \qquad (3)$$

$$T(x,t) = T_b, \qquad t = 0, \qquad 0 \leq x \leq L \qquad (4)$$

where c, k, w and $Q(x,t)$ are respectively the volumetric heat capacity (equal to the product of specific heat and density), thermal conductivity, the volumetric blood rate (equal to the product of the volumetric heat capacity and perfusion) and the volumetric heat source term. Initially the tissue domain is at uniform temperature T_b (blood temperature). For $t > 0$ the whole tissue domain is subjected to a space and time dependent heat power $Q(x,t)$ which represents the thermal source when using HIFU treatment technique [4]. Once the boundary and initial conditions, the thermal properties and the source term are specified, the direct problem defined by equations (1)-(4) can be solved analytically or numerically, and the temperature distribution can be calculated as a function of time and position anywhere in the considered tissue domain. Then the thermal dose is computed using Sapareto *et al.*'s model [5] and the quality treatment is checked with respect to the HIFU scheme and objectives.

Inverse problem

The inverse problem considered here is concerned with the estimation of three parameters defined by the following vector $U = [c, k, w]^T$, from the knowledge of transient temperature measurements taken at some locations inside the considered tissue domain. The transient temperature recordings are measured at N_s locations inside the tissue domain $0 < x_i < L$, $i = 1, \dots N_s$ where N_s designates the temperature sensor number.

As the specified boundary conditions (2)-(3) are of the first kind, the inverse problem theory [1, 6] requires that the measurements should be taken in this case inside the body, i.e., $0 < x_i < L$, $i = 1, \dots N_s$. When one or more boundary condition is of the second kind, temperature measurement can be collected at the boundary, i.e., $0 \leq x_i \leq L$, $i = 1, \dots N_s$. The inverse problem can then be regarded as an optimization problem which aims at finding the unknown thermal property vector $U = [c, k, w]^T = [u_1, u_2, u_3]^T$ (the superscript T denotes transpose). The basic process involves the minimization of the ordinary least squares norm given by:

$$S(U) = \sum_{i=1}^{N_s} \sum_{j=1}^{N_t} [T(x_i,t_j) - Y(x_i,t_j)]^2 \qquad (5)$$

where $T(x_i,t_j)$ and $Y(x_i,t_j)$ are respectively the computed and measured temperature at the sensor locations x_i, $i = 1, \dots N_s$ and different time t_j, $j = 1, \dots N_t$. Several minimization techniques can be applied to minimize the function (5) [1, 2, 7].

To determine the thermal properties of interest the function $S(U)$ is minimized with respect to each component of the vector U. This is accomplished by setting the first derivative of $S(U)$

with each parameter equal to zero, and solving for the estimated parameters U. The resulting expression, in vector form, for the estimated thermal properties is

$$U^{(k+1)} = U^{(k)} + \left[X^T X\right]^{-1} X^T \left[Y - T(U^{(k)})\right] \quad (6)$$

where the superscript (k) defines the iteration number and X represents the sensitivity matrix evaluated at the iteration (k). An iterative procedure is required even for linear conduction problems, due to the non-linear nature of the estimation problem because the coefficients of the sensitivity matrix depend on the unknown thermophysical properties to be recovered. Iteration continues until convergence of the estimated parameter is reached, i.e., when there is negligible change in any component of U. One criterion to indicate this is defined as:

$$\frac{| U^{(k+1)} - U^{(k)} |}{|U^{(k)}|} \leq \varepsilon \quad (7)$$

where ε is a small number to quantify convergence, such as 10^{-5}. The iterative procedure given by equation (6) is called the Gauss-Newton method [1], and requires the matrix $X^T X$ to be non-singular, or

$$\Delta = |X^T X| \neq 0 \quad (8)$$

where $| . |$ denotes the determinant. Equation (8) gives the so-called *Identifiability condition* [1]. In explicit form, the sensitivity matrix is given by :

$$X = \left[\frac{\partial T^T(U)}{\partial U}\right]^T = \begin{bmatrix} \frac{\partial T_1}{\partial u_1} & \frac{\partial T_1}{\partial u_2} & \frac{\partial T_1}{\partial u_3} \\ \frac{\partial T_2}{\partial u_1} & \frac{\partial T_2}{\partial u_2} & \frac{\partial T_2}{\partial u_3} \\ \vdots & \vdots & \vdots \\ \frac{\partial T_N}{\partial u_1} & \frac{\partial T_N}{\partial u_2} & \frac{\partial T_N}{\partial u_3} \end{bmatrix} \quad (9)$$

where $N = N_s \times N_t$. The sensitivity matrix coefficients can provide considerable insight to the estimation problem and in the design of the experiment for optimum accuracy in the estimates.

Results and discussion

In this section we present a set of numerical tests conducted by using the developed minimization algorithm. The first part of this section will be focused on an elementary sensitivity analysis. The second one concerns one numerical test case. The thermophysical properties of the tissue, presented in reference [8], are used in these simulations and given here. See table (1).

Table 1. Thermophysical properties of the analyzed tissue

	c	k	w	T_b
Value	3770000.	0.5	3770.	37.
Unit	$J/m^3 K$	W/mK	$W/m^3 K$	^oC

The heating source is taken to vary as time step variation only and is given by:

$$Q(t) = \begin{cases} 75000.\ W/m^3. & \text{for } 0.1 \times t_f \leq t \leq 0.9 \times t_f \\ 0. & \text{for otherwise} \end{cases} \quad (10)$$

The numerical simulations are performed for a one-dimensional heat conduction problem in a layer of thickness $L = 0.1\,m$ and over a time interval $t_f = 600.\,s$. The finite difference method using an uniform space grid and a purely implicit time scheme is used in the solution of this problem. A mesh grid with 101 (N_x) nodes in space and 101 (N_t) in time is used for all the results presented below. We prescribe the stopping criterion of 10^{-6} in expression (7).

Sensitivity analysis: Before presenting the estimation of the desired vector $U = [c, k, w]^T$, we present a short sensitivity analysis to show the feasibility (identifiability of the parameter set) of our estimation procedure. It is well known that an accurate parameter estimation can not be dissociated from an advanced sensitivity analysis in the goal to build the corresponding optimal experience scheme. The problem of choosing an optimal experimental design is usually conducted on the basis of the sensitivity coefficient study and the analysis of the D-optimum criterion [1, 2, 9]. The D-optimum criterion consists in the maximization of the determinant Δ of the square matrix $X^T X$ which results in minimizing the volume of the confidence region for the estimated parameters. This criterion is valid for additive and zero mean normal errors in the temperature measurements. By studying the timewise variation of the sensitivity coefficients and the D-optimum criterion prior to conducting experiments, such that equation (8) is maximized, the maximum information is available from an experiment.

The elements of the information matrix $X^T X$ are given by [1]:

$$X_{j,k} = \frac{1}{t_f} \sum_{i=1}^{N_s} \int_0^{t_f} \frac{\partial T(x_i,t)}{\partial u_j} \frac{\partial T(x_i,t)}{\partial u_k} dt \qquad j,k = 1,2,3 \quad (11)$$

where N_s is the number of temperature. This matrix characterizes the total sensitivity of the system at the measurements location x_i, to small variation of the entire set of parameters to be estimated.

First, we attempt to determine the minimal sensor N_s needed in this estimation procedure and the corresponding maximum criterion Δ. Then the found sensor number with their optimal locations x_i, $i = 1, \ldots N_s$ which corresponds to Δ_{max} are introduced in the estimation procedure to estimate the parameter of interest with the following initial guess of the unknown vector $U = [1.0000, 0.0001, 0.1000]^T$. The different steps of this procedure are summarized here:

1. We set $N_s = 1$
2. We compute the determinant $\Delta = |X^T X|$ at each space location using scanning method (*see the comment below*).
3. Then the found optimal position of the sensors is used in the minimization procedure with an initial guess of the unknown vector U.
4. If the estimation is successful, the optimal number of sensor is reached and their best locations are obtained.
5. Otherwise we increment $N_s = N_s + 1$ and return to step (2)

As the problem under investigation is symmetric, the search of optimal experimental factors is considered only over the half space domain $]0, L/2]$. To seek the optimal number of sensor and its coordinates we use a *scanning method* in the given spatial mesh: $M_x = N_x/2 + 1$. The search procedure consists in successive calculation and comparison of the values of the criterion Δ at the mesh nodes. For a single sensor $N_s = 1$, the procedure is reduced to compute Δ at each node x_i, $i = 2, \ldots M_x$ and to find the maximum value of Δ over the considered space region. When $N_s \geq 2$, let say $N_s = 3$ the first sensor is located at the first node after the boundary $x = 0$, which corresponds to x_2 in the spatial mesh grid, the second one is kept at the location x_3 while the third sensor is moved over the remaining nodes, i.e., x_i, $i = 4, \ldots M_x$. Then the first sensor is moved to x_3, the second one to x_4, and the third one over x_i, $i = 5, \ldots M_x$. This process is repeated until all spatial nodes were swept. To avoid a singular matrix $X^T X$, (purely numerically problem) two sensors are not allowed to hold at the same location. This means that sensor one will scan the nodes x_i, $i = 2, \ldots M_x - 2$. Sensor two will scan the nodes x_i, $i = 3, \ldots M_x - 1$. And sensor three will scan the nodes x_i, $i = 4, \ldots M_x$. At each combination the criterion Δ is computed and compared to the old one.

The results of searching optimal experimental parameters are presented in table (2). These results are obtained with the data mentioned at the beginning of this section. As the number of sensors is increased successively by one, initially ($N_s = 1$) there is a sharp increase in the optimum value of Δ and there are substantial changes in the optimum sensors coordinates. From $N_s = 2$ and greater the region for locating the sensors expands considerably and reaches its limit values ($x \simeq 0$ and $x = L/2$). Further increase in the number of sensors gives little increase in the criterion Δ, so any additional increase in that number is undesirable from the accuracy point of view. This result was also confirmed by the estimation procedure. Table (2) shows that any additional sensors more than two tend to have the same location as the first and the second sensor. This result is confirmed in the third case $N_s = 3$ and the small difference between x_2 and x_3 is due to the constraint that two sensors can not to be at the same location. A complete experimental optimization scheme should be conducted to identify the best conditions (optimal heating input, experiment duration, sensor locations, ...) for an accurate parameter estimation [9]. As a first conclusion here we can say that the minimal required number of sensors N_s and their optimal location are determined. In what follows, all the computations are performed with $N_s = 2$ and the corresponding optimal sensor locations given in table (2), i.e., $x_1 = 0.007\,m$ and $x_2 = 0.050\,m$.

Table 2. Maximal values for the criterion Δ and optimal sensor coordinates. Est. Proc.: estimation procedure. F: failed. C: converged. The number between brackets designates the iteration number required to obtained the parameters with the specified stopping criterion (7).

N_s	$\Delta_{max} \times 10^{24}$	x_1	x_2	x_3	x_4	Est.
1	10.03	0.009	-	-	-	F (11)
2	36701.41	0.007	0.050	-	-	C (23)
3	122205.45	0.006	0.049	0.050	-	C (23)
4	291593.62	0.006	0.007	0.049	0.050	C (23)

Sensitivity coefficients are very important because they indicate the change magnitude of the temperature $T(x,t)$ due to perturbations in the values of parameters of interest. They provide an indication of the importance of each parameter in the mathematical model, and they can also be used to determine whether or not the parameters are correlated, which is an important consideration in determining methodologies for the estimation of the parameters. Analytical expressions for the sensitivity coefficients can be obtained by taking the derivative of $T(x,t)$ in the direct problem (1)-(4) with respect to each unknown parameter. In our case a central finite difference approximation [1] is used to compute the sensitivity coefficients and they are denoted as:

$$X_c = c\frac{\partial T}{\partial c}, \qquad X_k = k\frac{\partial T}{\partial k}, \qquad X_w = w\frac{\partial T}{\partial w}$$

Figures (1), (2), and (3), present the time evolution of the sensitivity coefficients X_k, X_c, and X_w for the two considered position of sensors x_1, and x_2. As a first observation, the coeffi-

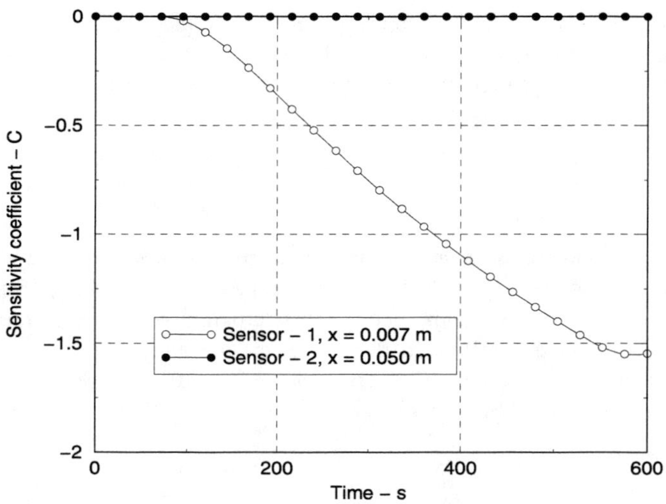

Figure 1. Evolution of sensitivity coefficients $X_k^{x_i}$ at different locations

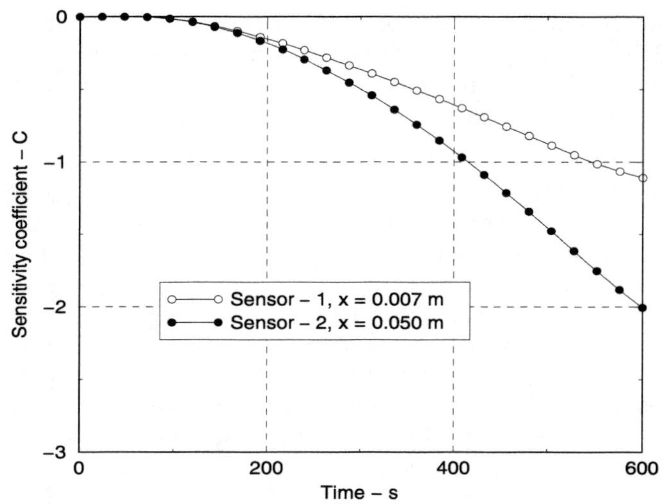

Figure 3. Evolution of sensitivity coefficients $X_w^{x_i}$ at different locations

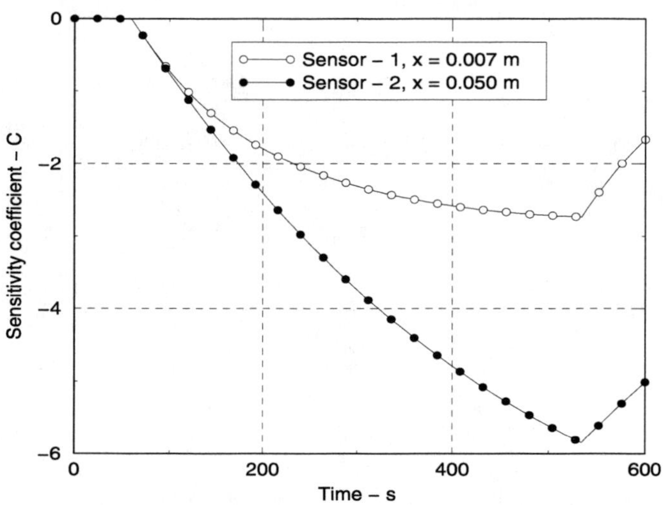

Figure 2. Evolution of sensitivity coefficients $X_c^{x_i}$ at different locations

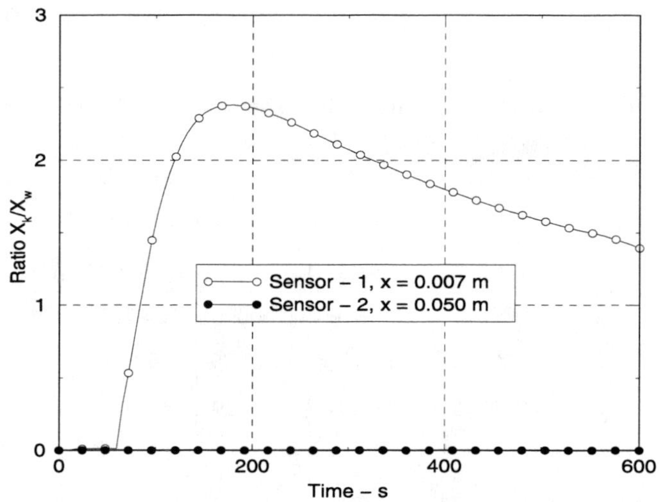

Figure 4. Evolution of the ratio $X_k^{x_i}/X_w^{x_i}$ at different locations

cients X_k and X_w seem to have the same time variation and could be correlated. This behavior can result in difficulties in estimating both of them at the same time. The inspection of the ratio between thermal conductivity k and the blood perfusion parameter w in terms of their respective sensitivity coefficients reveals that they don't present any correlation, see figure (4). Indeed, figure (4) shows clearly that the coefficients X_k and X_w are not linearly dependent and their simultaneously estimation can take place. A comparison of figure (2) and (1) shows that sensitivity coefficients with respect to c and with respect to k are not linearly dependent and both of them are negative during the treatment period. The biggest variation is observed in the case of X_c. The variation range is of the same order of temperature variation in the tissue. The lowest one is observed with the thermal conductivity parameter k at the second sensor location. For each sensor location, the shapes of the sensitivity coefficient curves are different. As expected the third sensor implemented in the middle of the studied domain presents the more pronounced variation for all parameters. After the heating period, the coefficient X_{cp} decreases while the coefficient X_k tends to a finite value. When the heating is stopped the coefficient X_c reacts instantaneously (in positive sense) because the occurrence of the cooling phenomenon into the heated tissue. In the view of this short sensitivity analysis we can state that the simultaneous estimation of volumetric heat capacity c, thermal conductivity k, and volumetric rate w is feasible.

Numerical test: The developed algorithm was first applied to the problem of estimating the vector $U = [c, k, w]^T$ without using noisy data. The stopping criterion used in the iterative process is given by equation (7), and the precision value is taken to be equal to: $\varepsilon = 10^{-6}$. Simulated temperatures obtained with the exact values of the vector U are used as measurements. The data used in this test are presented at the top of this section. Table (3) presents the obtained results by using the Gauss-Newton method starting from the following initial guess $U = [1.0000, 0.0001, 0.1000]^T$. After 23 iterations the parameter of interest are recovered with the desired precision.

Estimation error: To quantify the relative error of the estimation procedure, the following definition is introduced:

$$\varepsilon_p = \left| \frac{p - \bar{p}}{\bar{p}} \right| \times 100\% \qquad (12)$$

is the computation error for a given parameter p. The over-bar designates the exact parameter under hand.

Table 3. Recovered parameters using exact data

it	c	k	w	$S(U)$
1	1.000000	.000100	.100000	104.761 $\times 10^{12}$
2	2.000049	.000200	.200005	26.190 $\times 10^{12}$
3	4.000292	.003994	.400029	6.547 $\times 10^{12}$
⋮	⋮	⋮	⋮	⋮
21	3795483.53	.501914	3697.97	.1160
22	3769871.82	.499992	3770.60	.0000
23	3769999.99	.500000	3770.00	.0000
ε_p %	.0000	.0000	.0000	—

In a second stage we present the result of estimating $U = [c, k, w]^T$ when we use noisy data. The initial guess is taken the same as before. The simulated transient temperature data $Y(x,t)$ containing measurement errors are generated by adding random errors to the computed exact temperatures $T(x,t)$ as:

$$Y(x,t) = T(x,t) + \sigma\omega \qquad (13)$$

where σ is the standard deviation of measurement errors which is assumed to be the same for all measurements. For normally distributed random errors, there is a 99 % probability of the value of ω lying in the range:

$$-2.576 < \omega < +2.576 \qquad (14)$$

The values ω are generated randomly by the IMSL subroutine DRNNOR [10]. The value of σ in expression (13) was taken to be 0.05 (5.%). This results in perturbation of 0.3 % of the maximum temperature in the tissue reached at the end of one treatment cycle. Figure (5) displays the simulated temperatures and the applied heat source.

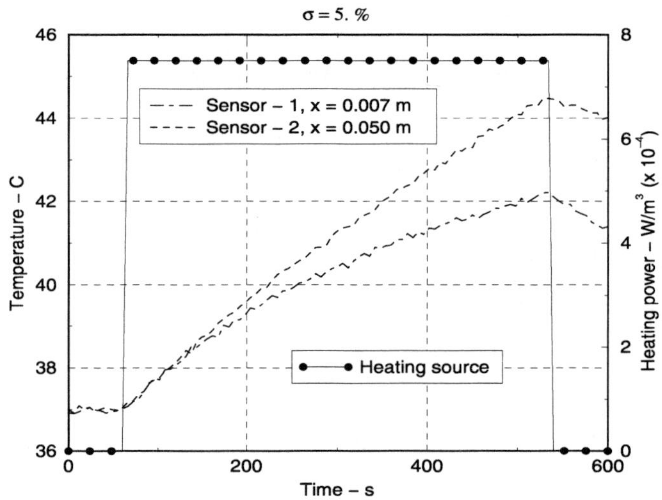

Figure 5. Simulated temperatures (noised date) and heat source

The results of estimating c, k, and w using the developed parameter estimation procedure and noisy data are reported in table (4). The three parameters are recovered in 23 iterations and the estimation error is less than 0.70 %. The highest estimation error is observed with c and w. The obtained results underline the efficiency of the developed algorithm and its capabilities to retrieve the exact parameters even with a poor initial guess.

Conclusion

This paper shows the application of the Gauss-Newton method to estimate simultaneously three thermo-physical parameters in hyperthermia problem. The presented results show that the method is valuable and efficient even in the absence of any a priori information on the initial guess of the parameters of interest. The study shows that the agreements between the estimated

Table 4. Recovered parameters using noised data

it	c	k	w	$S(U)$
1	1.000000	.000100	.100000	104.761×10^{12}
2	2.000049	.000200	.200005	26.190×10^{12}
3	4.000292	.003994	.400029	6.547×10^{12}
⋮	⋮	⋮	⋮	⋮
21	3791882.20	.498442	3722.58	3.1006
22	3766425.55	.496608	3794.71	2.9868
23	3766552.11	.496613	3794.12	2.9868
ε_p	.6773	.0915	.6399	—

and exact parameters are reasonable even when parameter estimation is performed with noisy data.

Acknowledgements

This material is based upon work supported by the National Science Foundation under Grant No. 9612044.

REFERENCES

[1] Beck J.V. and Arnold K.J. *Parameter Estimation in Engineering and Science*. Wiley Interscience, New York, 1977.

[2] Ozişik M.N. and Orlande H.R.B. *Inverse Heat Transfer : Fundamentals and Applications*. Taylor and Francis, Pennsylvania, 1999.

[3] Pennes H.H. Analysis of tissue and arterial blood temperatures in the resting human forearm. *J. Appl. Physiol.*, 1:93–122, 1948.

[4] Loulou T. and Scott E.P. Thermal dose optimization in high intensity focused ultrasound treatments using the adjoint method. *ASME-NHTC'00-12063*, August 20-22, Pittsburgh 2000.

[5] Sapareto S.A. and Dewey W.C. Thermal dose determination in cancer therapy. *Int. J. Radiation Oncology Biol. Phys.*, 10:787–800, 1984.

[6] Alifanov O.M., Artyukhin E.E. and Rumyantsev S.V. *Extreme Methods of Solving Ill-Posed Problems and t heirs Applications to Inverse Heat Transfer Problems*. Begell House, New York, 1995.

[7] Polak E. *Computational Methods in Optimization*. Academic Press, New York, 1971.

[8] Fan X. and Hynynen K. Control of the necrosed tissue volume during non-invasive ultrasound surgery using a 16 element phased array. *Med. Phys.*, 22(3):297–306, 1995.

[9] Taktak R., Beck J.V., and Scott E.P. Optimal experimental design for estimating thermal properties of composite materials. *Int. J. Heat Mass Transfer*, 36(12):2977–2986, 1993.

[10] IMSL. *Library Edition 10.0, User's Manual, Math/Library*. IMSL, 7500 Ballaire Blvd., Houston, Texas, 1987.

Proceedings of IMECE2002
ASME International Mechanical Engineering Congress & Exposition
November 17–22, 2002, New Orleans, Louisiana

IMECE2002-33647

MYOFIBROBLAST ACTIVITY IN AN ANIMAL MODEL FOR HUMAN HYPERTROPHIC SCAR

Derek A. Affonce
Mechanical Engineering Department
University of Massachusetts Dartmouth

Erik S. Sojka
Mechanical Engineering Department
University of Massachusetts Dartmouth

Livingston Van De Water
Center for Cell Biology and Cancer Research
Albany Medical College

Robert Sheridan
Division of Burns and Trauma
Massachusetts General Hospital
Shriners Burns Hospital

Alex J. Fowler
Mechanical Engineering Department
University of Massachusetts Dartmouth

INTRODUCTION

Approximately 100,000 burn patients require hospital admissions each year in the United States. About 90% of those patients survive to face the long term consequences of burn injury [1]. The primary cause of long term disability in burn survivors is hypertrophic scarring. These thick, deforming scars physically impair movement and cause major psychological morbidity. Hypertrophic scarring is particularly severe in young children [2].

There is currently no reliable treatment for the prevention or curing of hypertrophic scars [3]. One of the primary impediments to the development of new treatments is the complete lack of an animal model for hypertrophic scars. Since animals do not naturally exhibit hypertrophic scars there have been a number of attempts to create an animal model by transplanting human tissue onto athymic mice. It has been shown, for instance, that the hypertrophic scar tissue implanted into subcutaneous pockets in athymic mice will vascularize and grow over time [4].

Robb et al. [5] grafted both healthy human skin and hypertrophic scar tissue onto athymic mice. They found that the established scar was stable and retained its hypertrophic structure over time. The healthy skin was subjected to burn injury and it was found that it developed disordered dermal collagen, as is developed during hypertrophic scar formation. This animal model could provide a basis for testing new treatment modalities; but for a long time these results remained irreproducible.

Sojka et al. [6] reported that healthy human skin grafted onto athymic mice did develop hypertrophic morphology when burned under certain conditions. They found that this apparently hypertrophic response was very sensitive to the time and temperature of the burn. Polo et al. [7] established the long term viability and stability of established proliferative scars grafted onto athymic rats. They showed that proliferative scars of all types maintained high levels of human procollagen type III peptide (PIIP) production for up to six months following grafting onto a rat. High PIIP levels are characteristic of proliferative scar.

These two studies confirmed the findings of Robb et al. Polo et al.'s study also demonstrated that proliferative scars grafted onto athymic rats do maintain one of their disticintive physiological characteristics: hightened levels of human PIIP. Hypertrophic scars, however, also contain active myofibroblatss in the dermis that generate smooth muscle α actin. The sustained presense of dermal myofibroblasts is a feature unique to hypertrophic scars, and it has been proposed that it may be an alteration of myofibroblast disappearance that leads to hypertrophic pathology [8]. In order to prove that scars are hypertrophic in appearance and function, it is necessary to demonstrate the presense of active myofibroblasts in the dermis.

MATERIALS AND METHODS

All experiments were performed using NIH Swiss *nu/nu* mice (Taconic Farms, Germantown, NY) that were at least 8 weeks old. Human hypertrophic scar tissue obtained following surgical excision was cut to a size of approximately 3cm. by 2 cm. with a thickness of 3 mm. The sides of the tissue were then cut at about a 45° angle to shape the tissue into a flat topped pyramid. The angular cut along the edges of the tissue was used to increase the surface area of interaction between the thin mouse skin and the thick graft. Thicker and healthier grafts were achieved when using grafts with oblique rather than perpendicular sides.

Thirty-one mice received human hypertrophic scar transplants. Excision occurred at times ranging from two weeks to six months post transplantation. Samples from pre-graft controls and post excision were fixed in either 10% buffered formalin or 4% paraformaldehyde for embedding in paraffin or O.C.T. compound (Tissue-Tek, Sakura Finetek U.S.A., Inc., Torrance, CA) respectively. Samples embedded in paraffin were sectioned and stained with hematoxylin and eosin or Mason's Trichrome. Cryo sections were immunostained for detection of smooth muscle α actin.

Staining of smooth muscle muscle α actin was accomplished by using an anti smooth muscle α actin antibody that was grown up in a mouse, but was directly labeled with FITC. A test for non-specific binding was carried out by saturating the FITC labeled antibody with the peptide used to create it, thereby blocking all of the specific binding sites. Tissue samples incubated in a solution of antibody with preabsorbed (blocked) binding sites showed no fluorescence.

Figure 1. Fluorescent staining of smooth muscle α actin in the dermal tissue of an active hypertrophic scar graft 5 months after transplantation. Long lines of myofibroblasts laying down parallel lines of α actin are uniquely characteristic of hypertrophic scars.

RESULTS AND DISCUSSION

Human skin grafts healed to the mouse recipient bed and remained viable for up to six months. Microscopic examination of the human/mouse juncture showed a clear difference between the epidermis and the dermis of the mouse and human skin and immunological staining for mouse type I collagen showed that the dermal regions of the graft remained human in origin for the duration of the study [6]. As was reported by Robb et al. [5] we found that the scars retained the morphological characteristics of hypertrophic scar when studied with Mason's Trichrome and hematoxylin and eosin. The scars exhibited dense collagen nodules, abnormally high levels of vasculature and a large numbers of fibroblasts in structures that were essentially identical to those found in pre-graft controls. This hypertrophic morphology persisted for up to six months post-graft (which was as long as we tested).

Figure 1 shows an image of a hypertrophic scar excised five months after transplantation onto an athymic mouse and subsequently stained for smooth muscle α actin. Long lines of myofibroblasts laying down α actin are clearly visible. This activity is characteristic of active hypertrophic scars [8]. Samples collected at varying time periods for up to six months post graft, indicated that there was no reduction in myfibroblast activity over time after the tissue was grafted onto the mice. Some fresh control tissue did not show pre-graft myofibroblast acitivity; and grafts using those tissues did not show myofibroblast activity. For all thirty one grafts performed and tested we found that if myofibroblast activity was present in the pre-graft controls, then that activity was maintained after transplantation. Grafts that were originally inactive, remained inactive following transplantation. This proves that human hypertrophic scar that is grafted onto nude athmyic mice not only retains the morphological characteristics of hypertrophic scar, but also retains physiological activity that is uniquely characteristic of human hypertrophic scar.

We also checked for myofibroblast activity in healthy human skin that was grafted onto a mouse and subsequently subject to burn injury. We and others had previously reported that such skin could under certain circumstances develop hypertrophic morphology [5,6]; but the issue of its being truly hypertrophic in a physiological sense remained unresolved. Using immunofluorescent staining, however, we were able to show that there was myofibroblast activity in the regions that appeared to be hypertrophic. This not only lends weight to this model as being valid for the evaluation of scar treatments; but also suggests that the underlying cause of hypertrophy, which is unique to humans, may lie entirely within the skin tissue.

CONCLUSIONS

This paper demonstrates that human hypertrophic scar tissue that has been grafted onto an athymic mouse retains myofibroblast activity for up to six months. Since sustained myofibroblast activity is unique to hypertrophic scar and may be essential to its formation, its sustained presence in the scar grafts for long periods of time, lends significant weight to the argument that these grafts may provide a valid animal model for human hypertrophic scar. This model may provide a valuable tool for the evaluation and optimization of new treatment protocols.

We also showed that healthy human skin grafts on athymic mice that were subsequently burned could develop hypertrophic morphology and myofibroblast activity. It appears that these scars were truly hypertrophic. The creation of hypertrophic scar on human skin that is grafted onto a mouse, suggests that the cause of hypertrophy may lie entirely within the human skin itself, and not be due to some uniquely human metabolic response at a systemic level. While this issue clearly warrants further study, our results suggest that efforts to prevent hypetrophic scar should focus on the local response of the skin.

ACKNOWLEDGMENTS

This work was supported by a grant from the Whitaker Foundation.

REFERENCES

[1] Demling, R.H. and LaLonde, C. (1989) in **Burn Trauma**, eds. Blaisdell, F.W. and Trunkey, D.D., Thyme Medical Publishers, NY.

[2] Spurr, E.D. and Shakespeare, P.G. (1990), "Incidence of hypertrophic scarring in burn-injured children," Burns, v.16, pp. 79-81.

[3] Rockwell, W.B., Cohen, I.K. and Ehrlich, H.P. (1989), "Keloids and hypertrophic scars: a comprehensive review," Plastic and Reconstructive Surgery, v. 84, pp. 827-837.

[4] Kischer, C.W., Sheridan, D. and Pindur, J. (1989), "Use of nude (athymic) mice for the study of hypertrophic scars and keloids: vascular continuity between mouse and implants," The Anatomical Record, v. 225, pp. 189-196.

[5] Robb, E.C., Waymack, J.P., Warden, G.D., Nathan, P. and Alexander, J.W. (1987), "A new model for studying the development of human hypertrophic burn scar formation," JBCR, v. 8, pp. 371-365.

[6] Sojka, E. S., Balis, U., Sheridan, R., and Fowler, A.J. (1999), "Exploration of an animal model for hypertrophic scar formation," ASME HTD, v. 363, pp. 51-55.

[7] Polo, M., Kim, YJ., Kucukcelebi, A., Hayward, PG., Ko, F. and Robson, MC.(1998), "An in vivo model of human proliferative scar," Journal of Surgical Research, v. 74, pp. 187-195.

[8] Ehrlich, H.P., Desmouliere, A., Diegelman, R.F., Cohan, I.K., Compton, C.C., Garner, W.L., Kapanaci, Y. and Gabbiani, G (1994), "Morphological and immunochemical differences between keloid and hypertrophic scar," Am. J. Path. , v. 145, pp. 105-113.

Proceedings of IMECE2002
ASME International Mechanical Engineering Congress & Exposition
November 17–22, 2002, New Orleans, Louisiana

IMECE2002-33668

INVESTIGATION OF *IN SITU* THERMAL DENATURATION OF PROTEINS IN DUNNING AT-1 RAT PROSTATE CANCER CELLS USING FTIR AND DSC

Xiaoming He[1], Willem F. Wolkers[2], John H. Crowe[2], David J. Swanlund[1] and John C. Bischof[1]

[1]University of Minnesota, Department of Mechanical Engineering
[2]University of California at Davis, Section of Molecular and Cellular Biology

ABSTRACT

In this study, the *in situ* protein denaturation in Dunning AT-1 rat prostate cells was studied using FTIR and DSC. The denaturation information from FTIR showed a shift from α helix (amide-I band at 1655cm^{-1}) to β sheet (amide I band at 1620 cm^{-1}). The relative beta sheet area change between 20 °C and 70 °C was used to dynamically scale the thermal denaturation process during heating at 2 °C/min. DSC scans (at scanning rates of 2°C/min and 5°C/min) of the heat flow due to protein denaturation were recorded from 20 °C to 70 °C. The range over which protein denaturation occurs (from 45 °C to 70 °C) was consistent in both studies. By calorimetric experiments with DSC, the enthalpy change during protein denaturation was found to be 27.3±4.0J/g J/g and 25.9±0.5J/g, respectively. In addition, the activation energy and frequency factor were measured by kinetic experiments with both DSC and FTIR. A first order irreversible Arrhenius model was fit to the kinetic data using a flexible tolerance method. The activation energy (E) and frequency factor (A) between 20-70 °C was found to be 107.8kJ/mole and $3.4 \cdot 10^{14}$ 1/s for the FTIR study and 120.9kJ/mole and $3.6 \cdot 10^{16}$ 1/s for DSC analysis at a scanning rate of 2°C/min, respectively. At a scanning rate of 5°C/min, the DSC results gave an activation energy of 144.5kJ/mole and a frequency factor of $4.1 \cdot 10^{20}$ 1/s.

INTRODUCTION

Thermal therapy using minimally invasive probes is a surgical technique of growing interest and use in biomedicine. One of the key questions in the use of the technique is to understand the mechanism whereby the cells of the diseased tissue are destroyed. One implicated mechanism of thermal injury is protein denaturation [1]. Fourier Transform Infrared Spectroscopy (FTIR) is a powerful tool to study the overall protein secondary structure in cells, and hence can be used to detect changes in protein secondary structure associated with thermal injury. The amide-I and amide-II bands, arising from vibrations of protein backbones, have been widely used to determine the protein secondary structure of isolated proteins [2-5], and are diagnostic for the overall protein secondary structure of intact cells or organisms *in situ* [6]. Differential Scanning Calorimetry (DSC), which measures heat absorption due to thermally induced transitions [7], has been used to study the denaturation of pure proteins [8-9], erythrocyte membrane proteins [10-13], and overall protein denaturation in V79 cells [14-15] and hepatocytes [16]. When studying overall protein denaturation *in situ* in whole cells using DSC, identification of specific cellular proteins from the DSC curve is difficult, but this loss of specificity is offset by the ability to detect the overall transition of all proteins inside a cell.

This work reports the *in situ* protein denaturation during heating of AT-1 cells using both FTIR and DSC. The normalized excess calorimetric enthalpy (h_{cal}^{ex}) and area change over the amide-I band range between 1700 and 1600 cm^{-1} was used as a measure of the extent of protein denaturation. A first order irreversible kinetic model was fit to the experimental data to determine the kinetic parameters of thermal protein denaturation. Correlations between the results from these two techniques were also examined.

MATERIAL AND METHOD

AT-1 cell culture and preparation

AT-1 cells were propagated in DMEM/F-12 culture media supplemented with 5% fetal bovine serum (FBS), 1% penicillin/streptomycin in saline (Life Technologies, Gaithersburg, MD), 250 nM dexamethasone (Sigma-Aldrich, St. Louis, MO) and 5% CO_2 at 37°C. Cells were grown in 250 ml T flasks, lightly trypsinized with Trypsin-EDTA for 3 - 5 min and then collected. For FTIR study, the collected cells were pelleted at 50 x g for 2 min and resuspended in 1.0 ml of DMEM/F-12 with a final concentration of 1×10^6 cells/ml. The cells were then centrifuged at 14000 x g for 1 min, the medium was removed, and the cell pellet spread between two CaF_2 IR windows for FTIR analysis. For DSC study, the cells were pelleted at 400 x g for 10 min. The medium was removed and the cells were washed twice in degassed 1x D-PBS (Life Technologies, Gaithersburg, MD) at a volume ratio greater than 1:30. The cell pellet was then pipetted to a DSC pan for further DSC study. The concentration of protein in the cell pellet was measured to be 0.076±0.003mg/μl using bicinchoninic acid dye (BCA) reagents (Pierce Chemical Co., Rockford, IL).

FTIR

IR spectra were recorded on a Perkin-Elmer 2000 Fourier transform infrared spectrometer equipped with a liquid nitrogen-cooled Mercury/Cadmium/Telluride (MCT) detector using CaF_2 windows, as previously described [17]. In all cases, the AT-1 sample was cooled initially at 10°C/min from room temperature to -15°C using a Peltier temperature controller, and the sample temperature was monitored by a secondary thermocouple attached directly to the windows. The heating phase of the experiment was started immediately to prevent freezing of the sample, as verified by the spectral properties of the OH stretching and bending modes of water (3200 and 1650 cm^{-1}, respectively). Spectra were recorded over a temperature range from -15°C to +70°C at a heating rate of 2°C/min. Samples that are spontaneously nucleated were not used in this study. The spectral region from 1700 to 1600 cm^{-1} contains the amide-I absorption band of the protein backbones [4]. A spectrum of the buffer was subtracted from the cell spectra to eliminate the contribution of water to the amide-I region [18]. Heat denaturation profiles of intact cells were obtained by subtracting spectra recorded at 20°C from spectra at the indicated temperatures. Thermal denaturation of proteins was followed by monitoring the band between 1630 and 1610 cm^{-1}, and the area of this band was plotted as a function of temperature.

DSC

Differential scanning calorimetry profiles were obtained on a Perkin-Elmer Pyris 1 DSC. The DSC temperature calibration was done using pure water (0°C) and indium (156.5°C). The heat flow calibration was performed using indium (24J/g). The prepared samples at room temperature were loaded into the DSC holder at 20°C. The scan was started when the samples reached an equilibrated state. The scan was run to 70°C, the sample were then immediately cooled back to 20°C at a scanning rate of 200°C/min, held for 3 minutes and a second scan (rescan) run. The rescan was used to correct the first scan for baseline curvature and probable reversible transitions. Subtraction of the rescan from the first scan resulted in the heat absorption (Q^{ex}) due to any irreversible transitions involved. The excess specific heat (C_p^{ex}) was then calculated as follows:

$$C_p^{ex} = \frac{Q^{ex}}{m \cdot B} \quad (1)$$

where B is the scanning rate and m is the estimated weight of all proteins in each sample.

The Arrhenius kinetic model

By assuming the *in situ* protein denaturation in AT-1 cells during heating is a first order irreversible rate process, the fractional denaturation (F_d) of protein can be calculated using the Arrhenius model as follows:

$$F_d = 1 - \int_0^{\tau} A e^{-\frac{E}{RT}} dt \quad (2)$$

where t is time, τ is total heating time, $T = T_0 + Bt$, T_0 is the absolute temperature at t=0, B is the heating rate, A is the frequency factor, E is the activation energy, and R is the universal gas constant.

RESULTS AND DISCUSSION

FTIR

The IR spectra recorded at various elevated temperatures after subtracting the control spectrum taken at 20°C are shown in Fig. 1. By using difference spectra, one can avoid relying on other methods such as subtraction of the water band, which is subject to errors. The formation of extended β-sheet structures with heating can be deduced from the increase in absorbance of the bands at 1620 and 1695 cm^{-1}. The decrease in α-helical structures can be seen from the decrease in absorption around 1655 cm^{-1}. This shows that heat-induced protein denaturation coincides with a decrease in α-helical structures and a concomitant increase in extended β-sheet structures.

The area under the band between 1630 and 1610 cm^{-1} in Fig. 1 was calculated at different temperatures and normalized to area of the highest temperature (70±0.5°C). The average normalized area of three independent runs was plotted as a function of temperature in Fig. 2. This figure shows a slight increase in the β-sheet area, which likely reflects thermal rearrangements in protein secondary structure (changes in hydrogen bonding). These spectra also indicate that most proteins denatured above 45°C.

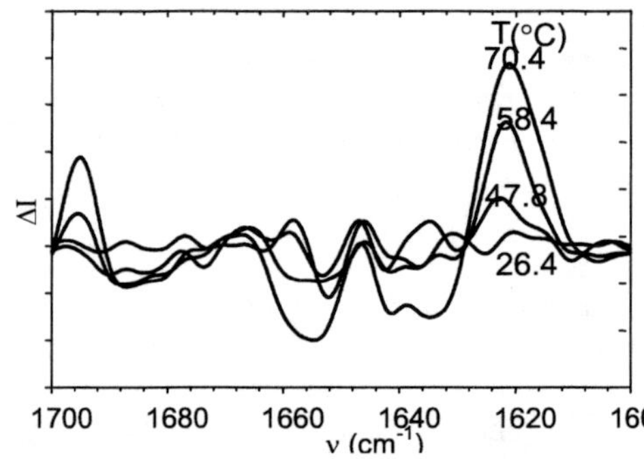

Fig. 1 Protein denaturation of AT-1 cells at various temperatures, showing difference spectra in the Amide-I region (difference intensity vs. wave number).

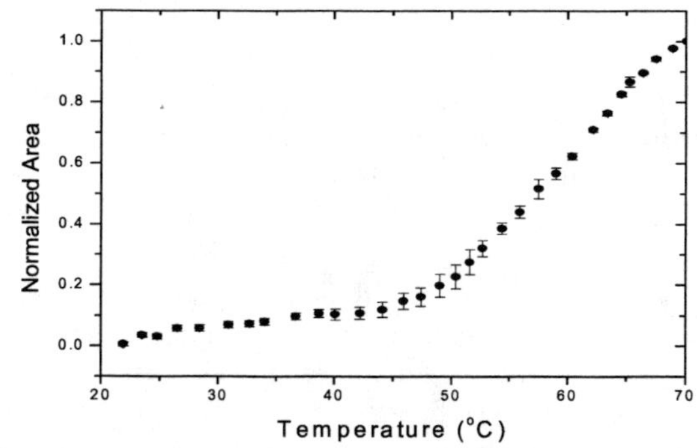

Fig. 2 Relationship between the normalized area between 1630 and 1610 cm^{-1} and temperature (at 2°C/min). The error bars represent SEM at each point.

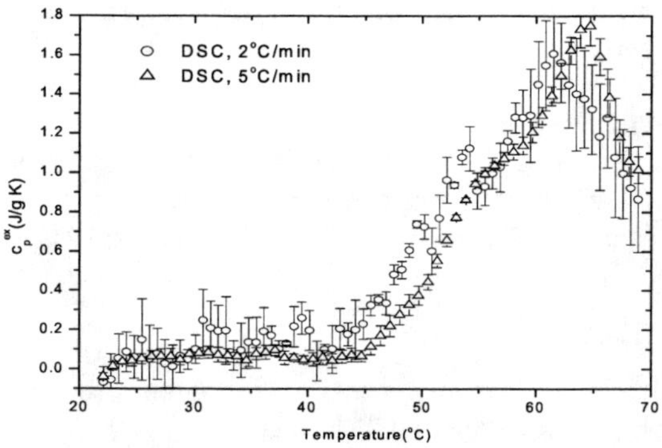

Fig. 3 Relationship between the excess specific heat and temperature at a scanning rate of 2°C/min. The error bars represent SEM at each point.

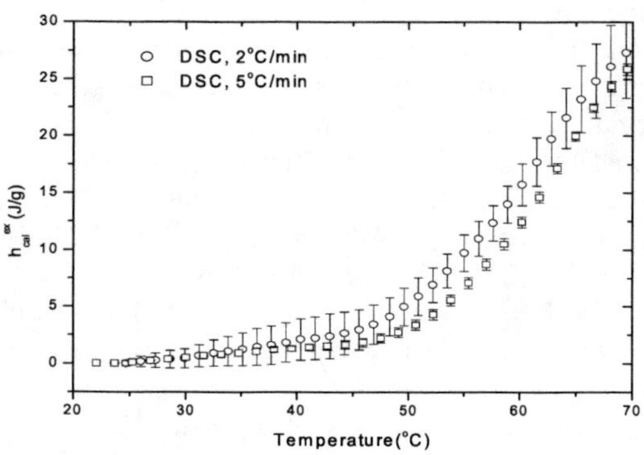

Fig. 4 Relationship between the excess calorimetric enthalpy and temperature.

DSC

The average excess specific heat (C_p^{ex}) of four independent runs is shown in Fig. 3. All intracellular transitions during DSC scanning will contribute to this endotherm, such as DNA and RNA unfolding, alterations of complex cellular structure, protein denaturation and denaturation-induced aggregation of macromolecules. However, most of the heat absorbed must be due to protein denaturation and contributions from all the other transitions should be small [13-16] as will be further corroborated later in this paper. Figure 3 also indicates that protein denaturation is not complete in AT-1 cells at 70°C because the excess specific heat is still high at that temperature.

The area under the curve of the excess specific heat represents the heat absorption due to protein denaturation (excess calorimetric enthalpy) and is given in Fig. 4. The overall excess calorimetric enthalpy after scanning to 70°C was found to be 27.3±4.0J/g and 25.9±0.5J/g for scanning rates of 2°C/min and 5°C/min, respectively. These data are a little lower but close to the average calorimetric enthalpy of 32 J/g when thermally denaturing pure proteins [8]. These values are also lower than that of 38±17J/g when thermally denaturing whole CHL V79 cells [14], where the scanning was stopped at a higher temperature (>100°C). Similar to Fig. 2, Fig. 4 shows a small increase of excess calorimetric enthalpy even before 37°C. This is probably due to thermal rearrangements of protein secondary structure.

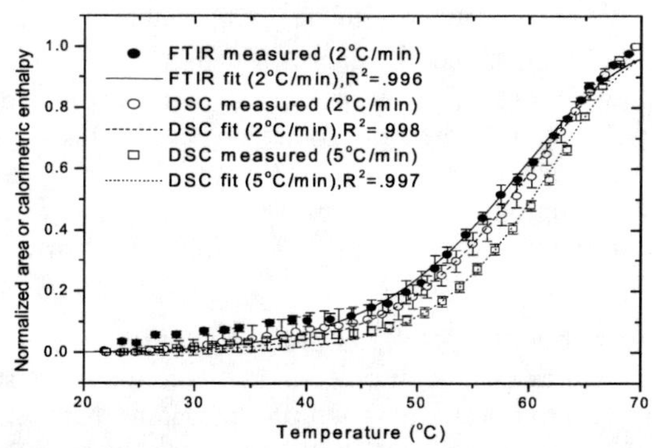

Fig. 5 Relationship between the normalized area and calorimetric enthalpy, and temperature. The error bars represent SEM at each point.

Protein denaturation kinetics

Figure 5 shows the normalized excess calorimetric enthalpy up to 70°C. The normalized area shown in Fig. 2 is also redrawn in this figure. Assuming the calorimetric enthalpy or area of denaturation of each protein in AT-1 cells is the same, the plot in Fig. 5 would be equal to the fractional denaturation of all cellular proteins. Since the heat absorption obtained from DSC scan or the area change obtained from FTIR of whole AT-1 cells is the summation of the individual transitions of all proteins, there will be an averaging such that the fractional enthalpy or area should be at least roughly approximate to the fractional protein denaturation [14].

In order to study the kinetics of protein denaturation quantitatively, the first order irreversible Arrhenius kinetic model (Eqn. 2) was fit to the experimental data in Fig. 5. This was done by minimizing the root mean square between the model fit and the experimental data using a flexible tolerance method [19-22]. The kinetic parameters (the activation energy, E and the fracquency factor, A) determined are given in Table 1.

Table 1 Kinetic parameters of *in situ* protein denaturation in AT-1 cells.

Parameters	FTIR (2°C/min)	DSC (2°C/min)	DSC (5°C/min)
E (kJ/mole)	107.8	120.9	144.5
A (s^{-1})	$3.4 \cdot 10^{14}$	$3.6 \cdot 10^{16}$	$4.1 \cdot 10^{20}$

As expected, the kinetic parameters of the fit to FTIR and DSC data at 2°C/min are very close to each other. Both are smaller than those of the fit to DSC data at 5°C/min, which suggests protein denaturation is dependent on both time and temperature. The activation energies in Table 1 are comparable to those of several pure proteins given in [20]. These activation energies lie between those reported for viability dye assays (244.8kJ/mole) and dynamic Calcein dye leakage (81kJ/mole) using the same cells [21]. Furthermore, these activation energies are close to those of an *in vivo* study of porcine kidney tissue heating with a microwave probe (around 98kJ/mole) [22]. However, they are much lower than that reported for clonogenics (526.39kJ/mole) using AT-1 cells [21].

As shown in Fig. 5, the measured data and fit results using the irreversible model match each other very well. This suggests that the protein denaturation captured by both FTIR and DSC in this study is approximately an irreversible process, and the contribution from any reversible transitions, if they exist, should be small.

Correlations between the FTIR and DSC study

Since the area change during heating from the FTIR study is only due to protein denaturation, the proximity between the DSC and FTIR measurements at 2°C/min (as shown in Fig. 5 and Table 1) suggests that the measured enthalpic change with temperature using DSC is primarily due to protein denaturation. At the same heating rate (2°C/min), Fig. 5 shows that the normalized excess enthalpy up to 70°C lies slightly below the normalized area from the FTIR study in the whole temperature range. This difference may be due to the difference in sensitivity of the two instruments.

SUMMARY

The *in situ* thermal denaturation of Dunning AT-1 prostate tumor cells was studied using FTIR and DSC. FTIR study showed that protein denaturation coincided with a decrease in α-helical structures and a concomitant increase in extended β-sheet structures. By subtracting the rescan from the first scan, DSC analysis gave a transition enthalpy of 27.3±4.0J/g (2°C/min) and 25.9±0.5J/g (5°C/min) from 20°C to 70°C. The proximity between the activation energies from DSC and FTIR at the same heating rate indicates that the heat flow captured by DSC is mainly due to protein denaturation. The good match between measured data and fit results from the first order irreversible kinetic model suggests that the protein denaturation measured by both DSC and FTIR is primarily an irreversible process. A first order irreversible Arrhenius model was fit to the kinetic data using a flexible tolerance method. The activation energy (E) and frequency factor (A) between 20-70 °C was found to be 107.8kJ/mole and $3.4 \cdot 10^{14}$ 1/s for the FTIR study and 120.9kJ/mole and $3.6 \cdot 10^{16}$ 1/s for DSC analysis at a scanning rate of 2°C/min, respectively. At a scanning rate of 5°C/min, the DSC results gave an activation energy of 144.5kJ/mole and a frequency factor of $4.1 \cdot 10^{20}$ 1/s. These kinetic parameters can be used to predict the dynamic *in situ* protein denaturation in AT-1 cells in the temperature range described herein.

ACKNOWLEDGMENTS

Thanks to Urologix, Inc. (Plymouth, MN) for their support of this study. This work was also supported partially by grants HL57810 and HL61204 from the NIH, and 981711 from DARPA.

REFERENCES

1. Lepock, J. R., "Involvement of membranes in cellular responses to hyperthermia", Radiat. Res., Vol. 92, 1982, pp433-438
2. Bandekar, J., "Amide modes and protein conformation", Biochimica et Biophysica Acta, Vol. 1120, 1992, pp123-143
3. Byler, D. M., and Susi, H., "Examination of the secondary structure of proteins by deconvolved FTIR spectra", Biopolymers, Vol. 25, 1986, pp469-487
4. Surewicz, W. K., Mantsch, H. H., and Chapman, D., "Determination of protein secondary structure by Fourier transform infrared spectroscopy: A critical assessment", Biochemistry, Vol. 32, 1993, pp389-394.
5. Susi, H., Timasheff, S. N., and Stevens, L., "Infrared spectra and protein conformations in aqueous solutions. The amide I band in H_2O and D_2O solutions", Journal of Biological Chemistry, Vol. 242, 1967, pp5460-5466
6. Wolkers, W. F., Alberda, M., Koornneef, M., and Hoekstra, F. A., "Heat stability of proteins in maturation defective mutants of *Arabidopsis thalania*: A FT-IR microspectroscopy study", Plant Journal, Vol. 16, 1998, pp133-143
7. Höhne, G. W. H., Hemminger, W. and Flammersheim, H.-J., "Differential scanning calorimetry – an introduction to practioners", Springer - Verlag Berlin, Heidelberg, Germany, 1996

8. Privalov, P. L. and Khechinashvili, N. N., “A thermodynamic approach to the problem of stabilization of globular protein structure: a calorimetric study”, J. Mol. Biol., Vol. 86, 1974, pp665-684
9. Miles, C. A., Burjanadze, T. V. and Bailey A. J., “The kinetics of the thermal denaturation of collagen in unrestrained rat tail tendon determined by differential scanning calorimetry”, J. Mol. Biol., Vol. 245, 1995, pp437-446
10. Jackson, W., Kostyla, J., Nordin, J. and Brandts, J. F., “Calorimetric studies of protein transitions in human erythrocyte ghosts”, Biochemistry, Vol. 12, 1973, pp3662-7
11. Brandts, J. F.,Reickson, L.,Lysko, K., Schwartz, A. T. and Taverna, R. D., “Calorimetric studies of the structural transitions of human erythrocyte membrane: the involvement of spectrin in a transition”, Biochemistry, Vol. 16, 1977, pp3450-54
12. Lysko, K. A., Carlson, R.,Taverna, R., Snow, J. and Brandts, J., “Protein involvement in structural transitions of erythrocyte ghosts: use of thermal gel analysis to detect protein aggregation”, Biochemistry, Vol. 20, 1981, pp5570-76
13. Lepock, J. R., Frey, H. E., Bayne, H. and Markus J., “Relationship of hyperthermia-induced hemolysis of human erythrocytes to the thermal denaturation of membrane proteins”, Biochimica et Biophysica Acta, Vol. 980, 1989, pp191-201
14. Lepock, J. R., Frey, H. E., Rodahl, A. M. and Kruuv J., “Thermal analysis of CHL v79 cells using differential scanning calorimetry: implication for hyperthermic cell killing and the heat shock response”, Journal of cellular physiology, Vol. 137, 1988, pp14-24
15. McDuffee, A. T., Senisterra, G., Huntley, S., Lepock, J. R., Sekhar, K. R., Meredith, M. J., Borrelli, M. J., Morrow, J. D. and Freeman, M. L., “Proteins containing non-native disulfide bonds generated by oxidative stress can act as signals for the induction of the heat shock response”, Journal of cellular physiology, Vol. 171, 1997, pp143-151
16. Lepock. J. R., Frey, H. E. and Ritchie, K. P., “Protein denaturation in intact hepatocytes and isolated cellular organelles during heat shock”, Journal of Cell Biology, Vol. 122, 1993, pp1267-76
17. Wolkers, W. F., Walker, N. J., Tablin, F. and Crowe, J. H., “Human platelets loaded with trehalose survive freeze-drying”, Cryobiology, Vol. 42, 2001, pp79-87
18. Haris, P. I., Coke, M., and Chapman, D., “Fourier transform infrared spectroscopic investigation of rhodopsin structure and its comparison with bacteriorhodopsin”, Biochim. Biophys. Acta, Vol. 995,1989, pp160-167
19. Himmelblau, D. M., “Applied nonlinear programming”, McGraw- Hill Inc., 1972
20. Johnson, F. H., Eyring, H. and stoner, B. J., “Temperature”, Chapter 3 in *The theory of rate process in biology and medicine*, John Willey & sons, New York, 1974
21. Bhowmick, S., Swanlund, D. J. and Bischof, J. C., “Thermal therapy of Dunning AT-1 prostate tumor cells”, ASME JBME, Vol. 122, 2000, pp51-59
22. He, X., Mcgee, S., Coad, J. E., Swanlund, D. J., Rudie, E., Kluge, S. and Bischof, J. C., “Investigation of the thermal and injury behavior during microwave thermal therapy of porcine kidney”, submitted to section of heat transfer in medicine and biology, IMECE 2002

Proceedings of IMECE2002
ASME International Mechanical Engineering Congress & Exposition
November 17–22, 2002, New Orleans, Louisiana

IMECE2002-33678

HEAT SHOCK PROTEIN 70 EXPRESSION KINETICS

Sihong Wang, Kenneth R. Diller, Shanti J. Aggarwal
Department of Biomedical Engineering
University of Texas at Austin, Austin, TX 78712, USA

ABSTRACT

HSP70 is well known for its major role in cardiac ischemia protection. The purpose of this study was to determine the HSP70 expression kinetics for new protocol design in cardiac surgery, based on HSP70 protection function in clinical applications. Bovine aortic endothelial cells (BAEC) were used in experiments. Cells were heated at 42°C at different time intervals up to 5 hours and subsequently incubated at 37°C for up to 48 hours. Western blot and quantitative protein analysis were performed to measure HSP70 expression. The expression kinetics is a function of thermal stress time as well as post-stress time. At least three stages were identified for the kinetics curve: increasing, maximum plateau and decreasing regions. The peak HSP70 concentration is 10 times the basal level for western blot analysis in BAECs. Two hours incubator heating followed by twelve hours post-heating falls in the plateau region. This research result provides information applicable to evaluation of energy sources and heating methods to induce optimal HSP70 expression in a target tissue.

INTRODUCTION

Heat shock proteins (HSPs) are molecular chaperones. They protect cells from extreme physiological, pathological and environmental conditions [1]. The cytoprotective functions of HSPs have been found in many organs, such as heart [2] and brain [3]. Transient exposure to elevated temperature or some non-toxic chemicals gives a protective effect against long-term exposure to lethal stresses due to the increased HSP expression in tissue. HSP70, with molecular weight 70 kDa, is known to be the major molecular chaperone in all the cellular compartments and organs. Whole body hyperthermia [4] or local heating [5] of a heart increases the myocardial HSP70 level and protects against ischemia-reperfusion injury following cardiac surgery.

A major limit in cardiac surgery and transplantation is the accepted safe ischaemic time during the surgery, which still remains 4 hours. Over-expression of HSP70 has a great potential to extend this time. Optimal HSP70 expression is critical to reach the maximum level of myocardial protection [5]. Therefore it is important to control the thermal stress regimen to produce an optimal amount of expression. When the maximum HSP70 expression appears and how long it can be sustained is also critical to the design of improved cardiac surgical procedures. A complete understanding of temperature and time kinetics of HSP70 expression is necessary to determine the best protocol inducing protection. Currently comprehensive HSP70 expression kinetics data for heart tissue or cardiac cells is not available.

Bovine aortic endothelial cells (BAEC) were chosen for study as representative of cardiac tissue. Endogenous HSP70 kinetics were effected and measured by incubator heating and western blot analysis. A large range of heating time and post-heating time combinations were examined during the experiments. This data for the expression kinetics provides an effective guide to the design of heating protocols for cardiac surgery involving HSP70 protection.

MATERIALS AND METHODS

BAECs were grown in 10cm culture dishes until they were 90% confluent. Cells were heated in a 42°C incubator and then allowed to recover subsequently in a 37°C incubator, both with 5% CO_2. The heating time was measured from when cells were transferred into 42°C incubator to the time when cells were removed. Cells were separated into 14 groups according to the different recovery times. A matrix of heating and recovery times was used: heating times varied from 30 minutes to 5 hours (0, 0.5, 1, 1.5, 2, 3, 4, 5 hours) and recovery times varied from 0 to 48 hours (0, 1, 2, 3, 4, 5, 6, 7, 8, 10, 12, 15, 24, 48 hours). Cells were lysed immediately after heating and recovery with lysis buffer, and supernatant solutions containing HSP70 were saved for further protein analysis. Western blot was used to determine the expression of inducible HSP70 in all samples. To assess the temperature effect on the HSP70 expression, cells were heated at temperatures from 40 to 44°C for two hours with a uniform recovery period of 16 hours. HSP70 expression, which is represented by the ratio of HSP70 to actin, is further normalized to the HSP70-actin ratio of the control.

RESULTS

Figure 1 shows that HSP70 expression is a function of heating/stress time and post-heating/post-stress time. The maximum HSP70 expression of ten times larger the basal level occurs for around 2 hours heating followed by 12 hours post-heating.

Figure 2 shows the HSP70 expression of varied heating times followed by 12 hours post-heating. For fixed 12 hours post-heating samples, heating time longer than 2 hours does not make a difference with respect to HSP70 expression. The data indicates that there may exist a threshold of thermal dose for maximum HSP70 expression.

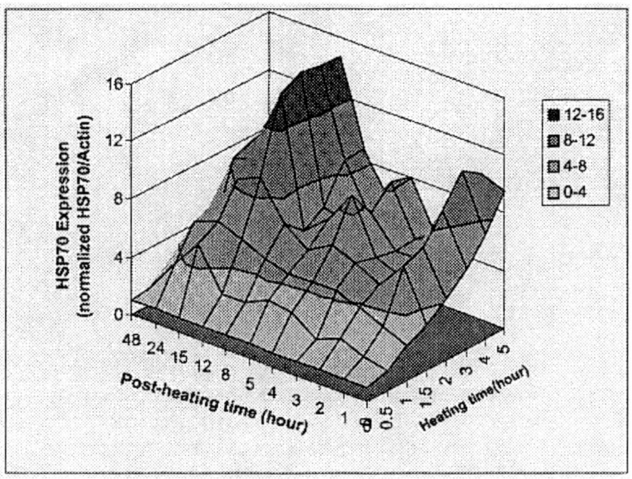

Figure 1. HSP70 expression kinetics

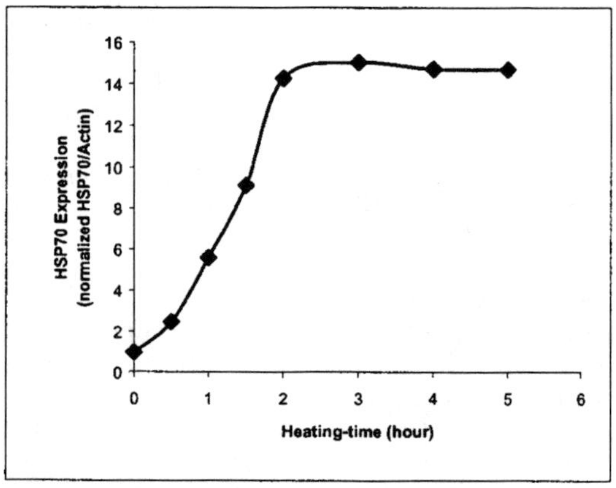

Figure 2. HSP70 expression over different heating times followed by 12 hours post-heating

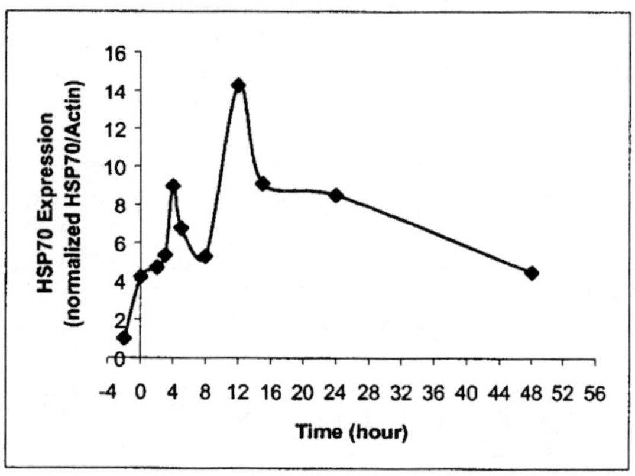

Figure 3. HSP70 expression at 2 hours heating followed by up to 48 hours post-heating

There are at least three stages in the HSP70 kinetic expression curve: increase, plateau/peak and decrease (Fig. 1). Figure 3 shows a two-peak phenomenon when the HSP70 expression was plotted as a function of the post-heating time.

Thermal stress triggers HSP70 expression. After the thermal stress is removed, HSP70 expression continues until it reaches the plateau after 10 hours post-heating, then it stays at the plateau for several hours, followed by a drop. At 24 hours after the thermal stress, HSP70 concentration is still more than 6 times larger than the control/basal level. Temperatures from 42°C to 43°C induce a higher level of HSP70 expression than lower temperatures.

CONCLUSION AND DISCUSSION

HSP70 expression kinetics is a function of at least three parameters: heating temperature and time, and post-heating/recovery time. There are at least three stages in the kinetics curve of HSP70 induction: increasing, plateau and decreasing regions. Judging by the western blot analysis, the peak HSP70 concentration is 10 times larger than the basal level. Two hours incubator heating followed by 10 to 16 hours post-heating will induce the maximum HSP70 expression. Post-heating time plays an important role in the HSP70 expression kinetics. The two-peak phenomenon in the post-heating region of the HSP70 expression is consistent with the "second-window protection phase" from ischemic preconditioning analysis [6].

The saturated HSP70 expression at over 2 hours incubator heating and 12 hours post-heating indicates that a threshold of thermal doses exists for maximum HSP70 expression. On the other hand, the different HSP70 expression levels at different temperatures for the fixed heating and post-heating time means that the rate of thermal input may be another factor that affects the HSP70 expression kinetics. Alternative strategies for more intense heating for shorter time periods should be investigated further. We are developing a perfusion heating stage with a culture dish which reduces the heating time ramp to less than 5 minutes and provides the same *in vitro* environment as the incubator to conduct additional investigations.

ACKNOWLEDGMENTS

This work is supported by Abell-Hanger Foundation.

REFERENCE

1. Kiang, J. G., and Tsokos, G. C., 1998, "Heat Shock Protein 70kDa: Molecular Biology, Biochemistry, and Physiology," Pharmacol. Ther. **80**, No. 2, pp. 183-201.
2. Gray, C. C., Yacoub, M. H., et al, 1999, "Heat Stress Proteins and Myocardial Protection: Experimental Model or Potential Clinical Tool?" The International J. of Biochemistry & Cell Biology, **31**, pp. 559-573.
3. Yenari, M. A., Steinberg, G. K., et al, 1999, "The Neuroprotective Potential of Heat Shock Protein 70 (HSP70)," Molecular Medicine Today, **5**, pp. 525-531.
4. Currie, R. W., Mailer, K., et al, 1988, "Heat Shock Response is associated with Enhanced Postischemic Ventricular Recovery," Circ. Res. **63**, pp. 543-549.
5. Gowda, A., Motamedi, M., et al, 1998, "Cardioprotection by Local Heating: Improved Myocardial Salvage after Ischemia and Reperfusion," Ann. Thorac. Surg., **65**, pp. 1241-1247.
6. Yellon, D. M., Baxter, G. F., 1995, "A 'second window of protection' or Delayed Preconditioning Phenomenon: Future Horizons for Myocardial Protection?" J. Thorac. Cardiovasc. Surg. **115**, pp. 200-209.

Proceedings of IMECE2002
ASME International Mechanical Engineering Congress and Exposition
November 17-22, 2002, New Orleans, LA

IMECE2002-33684

THERMOSTABILITY STUDIES OF DESICCATED MURINE SPERMATOZOA NUCLEAR DNA TO PREDICT THE BENEFICIAL EFFECT OF TREHALOSE IN LONG TERM STORAGE

Sankha Bhowmick, Bharat D. Nath
Center for Engineering In Medicine,
Massachusetts General Hospital,
Harvard Medical School and Shriners
Hospital for Children, Boston, MA.

John D. Biggers
Dept. of Cell Biology
Harvard Medical School
Boston, MA.

Mehmet Toner
Center for Engineering In Medicine,
Massachusetts General Hospital,
Harvard Medical School and Shriners
Hospital for Children, Boston, MA.

ABSTRACT
Long term preservation of mouse sperm in a desiccated state using sugars like trehalose may offer attractive economic benefits in the management of rapidly increasing transgenic mouse strains. The goal of the current study was to evaluate the protective effect of intracellular trehalose on sperm nucleus by predicting the long-term nuclear degradation kinetics of desiccated spermatozoa using an Arrhenius model whose parameters are obtained from high temperature-short time storage studies. B6D2F1 sperm isolated in an EGTA supplemented tris-HCl buffer (with or without 0.5M intracellular trehalose) were convectively dried with inert nitrogen gas in a controlled manner to moisture content <5%. The samples were then vacuum packed and stored at 22, 37, 45, 60 and 90°C for 1, 3 or 7 days. Following rehydration, the sperm sample was assayed for DNA damage using the sperm chromatin structure assay (SCSA). Results indicate significantly ($p<0.05$) lower DNA degradation for cells dried with intracellular trehalose at 45, 60 and 90°C for 1, 3 or 7 days compared to cells dried without trehalose. Based on a 10% increase in the index of injury, the calculated activation energy and frequency factors were 10.33 kcal/mole and 5.4×10^5 hr^{-1} respectively for cells dried in EGTA solution only. The corresponding numbers for cells dried in EGTA solution supplemented with 0.5M trehalose were 5.7 kcal/mole and 43.73 hr^{-1}. Based on these parameters the time required for 10% DNA degradation are 279 and 759 hours for samples desiccated in plain EGTA vs. trehalose supplemented EGTA. These results indicate the beneficial effect of intracellular trehalose for the long-term storage of desiccated sperm.

KEYWORDS: desiccation, mouse sperm, trehalose, thermostability, nuclear DNA

INTRODUCTION
Desiccation offers an attractive alternative to cryopreservation by offering the possibility of long-term storage at or near ambient temperatures. In nature, many anhydrobiotic organisms survive desiccation by imbibing large quantities of sugar (Crowe et al., 1992). Sugars have also been shown to be beneficial in the desiccation of various biologics including proteins and membranes (Carpenter et al., 1987; Crowe et al., 1987). Although, there have been studies on desiccation of mammalian cells using sugars, their long-term storage potential has not been demonstrated in these studies (Puhlev et al., 2001). One possible high impact area for desiccation is the maintenance of transgenic mouse sperm libraries. For mouse sperm, the stability of nucleus is the primary requirement for successful production of offsprings (Wakayama and Yanagimachi, 1998). Kusakabe et al. (2001) used an EGTA solution to successfully freeze-dry various strains of mouse sperm, store them for about 3 months, and subsequently rehydrate them and successfully form fetuses after ICSI into oocytes. However, the effect of sugars on the storage of mouse sperm after desiccation has not been studied.

In an attempt to elucidate the possible beneficial effect of sugars in the long-term storage of mouse sperm, we desiccated mouse sperm in EGTA solution supplemented with or without trehalose and stored them at different temperatures at and above ambient for a short period of time (1,3 and 7 days). We then used the SCSA to quantitatively determine the DNA damage and used the Arrhenius damage model to show that there is an improvement in the storage stability of sperm nuclear material when desiccated and stored in EGTA supplemented with trehalose compared to EGTA solution alone.

MATERIALS AND METHODS

Isolation of Sperm and Loading Trehalose: 9-14 weeks old male B6D2F1 mice (Jackson Laboratories, Bar Harbor, ME) were euthanized, and their epididymis collected in 0.5 ml of EGTA solution (Kusakabe et al., 2001). It was subsequently punctured with a 28-gauge needle tip and the solution was incubated at 37°C for 10 minutes to allow dispersion of the sperm. 0.4 ml sperm containing EGTA solution was aspirated and used for experiments. For loading of trehalose, 50 µl of a 135 µg/ml solution of H-5 solution (Chen et al., 2001) was mixed with 200 µl of the sperm-EGTA solution to give a concentration of ~25 µg/ ml. The solution was incubated at ambient temperature for 15 minutes after which a vital dye uptake study showed > 90% poration. The solution was then loaded with an equal volume of a 1M-trehalose solution. The trehalose loaded sperm solution was allowed to incubate at ambient temperature for 45 minutes prior to desiccation.

Desiccation and Storage: The desiccation protocol was performed using a rapid convective drying protocol as outlined previously (Bhowmick et al., 2002). 20 µl of sperm solution was placed on an antibody-staining slide (Gold Seal, Portsmouth, NH) and placed in a chamber, tightened to prevent leakage and dried with a steady flow of nitrogen gas at 5 liters per minute. The drying kinetics were such that the EGTA solution reached <5% moisture in 10 minutes and the trehalose solution reached the same moisture content in 15 minutes. Moisture content was measured by gravimetric analysis. The samples were then vacuum packed and stored at 22, 37, 45, 60 and 90°C for 1, 3 or 7 days. After storage, the samples were rehydrated to their original volume by adding deionized water and used for SCSA.

SCSA: The SCSA utilizes the unique property of acridine orange which emits red fluorescence when it binds to single stranded DNA, while emitting green fluorescence when binding to double stranded DNA. Since sperm nucleus is devoid of RNA, the SCSA can be used to detect the fraction of fragmented DNA. Details of the SCSA procedure and analysis are described elsewhere (Evenson and Jost, 2000). Briefly, the desiccated sperm is rehydrated to a final volume of 100 µl. 200 µl of acid/detergent buffer was added followed by addition of 600 µl of AO staining solution after 30 seconds. The solution is equilibrated for 2 minutes after which it is run through a flow cytometer at 100-200 particles/s. The entire procedure after rehydration is performed at 4°C. The SCSA was used to quantify the index of injury, defined as the ratio of the transformed geometric mean of the red fluorescence to the total fluorescence (red+green). The index of injury for each temperature-time point and their normalized change with respect to control enabled the determination of Arrhenius parameters for the different conditions of drying.

Arrhenius Damage Model: The Arrhenius damage model is based on assumption that thermal damage can be assumed to follow the first order reaction rate kinetics and was first espoused by Henriques (1947). The equation is expressed as:

$$\frac{d\Omega}{dt} = \frac{\Delta\Omega}{\Delta t} = \frac{1}{\tau} = Ae^{-\frac{E}{RT}} \qquad (1)$$

In Eq. 1, Ω is the assay measured time dependant accumulated damage; t is the time for damage accumulation; $\Delta\Omega$ equals one and represents the assay dependant threshold of damage, τ is the time required to accumulate irreversible damage; E is the activation energy, A is the frequency factor; T is the absolute temperature and R is the universal gas constant.

RESULTS AND DISCUSSION

Fig. 1 demonstrates an increasing degradation of sperm nuclear DNA desiccated in EGTA compared to samples desiccated in EGTA-trehalose solution and stored at higher temperatures. The y-axis represents the normalized index of injury (baseline value of 100). At 22 °C, there is no significant difference between samples stored for 1, 3 or 7 days after drying in either

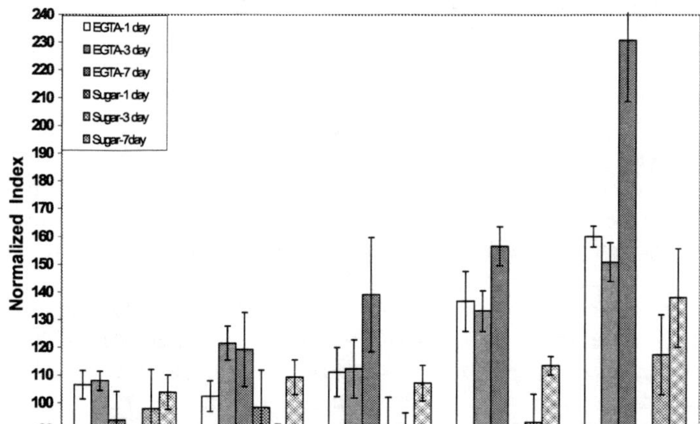

Fig. 1 Plot of normalized index of injury (calculated based on the ratio of the transformed geometric mean of the red fluorescence and the total fluorescence) at various storage temperatures. An increase in the index implies a larger degradation of sperm nuclear DNA (n=4).

EGTA or trehalose solution. At 37°C, the EGTA dried samples stored for 3 and 7 days show an increase in DNA degradation compared to their corresponding counterpart dried in trehalose, but these results are not significantly different from control. Significant increase in DNA degradation compared to control ($p<0.05$) was found for samples dried in EGTA solution and stored at 45°C for 7 days and each of 1, 3 and 7 days at 60 and 90°C. At 45, 60 and 90°C, all samples dried in EGTA solution show significantly more ($p<0.05$) DNA degradation compared to samples dried in trehalose solution. Despite a trend of increasing DNA degradation with increasing temperatures and time of storage after drying in trehalose solution, the results for all points except 90°C- 3 and 7 days storage are not significantly different from control indicating a high degree of protective effect to the sperm nuclear DNA in the presence of trehalose.

In order to predict the long-term storage capability at ambient conditions, the Arrhenius damage model was used to evaluate the activation energy (E) and the frequency factor (A) using a 10% increase in normalized index of injury as the threshold of DNA degradation. Since significant difference in DNA degradation was obtained in the 45-90°C range for storage conditions up to 7 days, these data points were used for the calculation of the parameters. Based on a 10% increase in the index of injury, the calculated activation energy and frequency factors were 5.7 kcal/mole and 43.73 hr^{-1} and 10.33 kcal/mole and $5.4x10^5$ hr^{-1} respectively for cells stored with and without trehalose respectively. The corresponding computed times for 10% DNA degradation at 22°C are 759 and 279 hours respectively, demonstrating that sperm can be stored after drying in trehalose supplemented EGTA solution for a longer period of time with minimal DNA damage compared to plain EGTA solution.

In summary, the current study demonstrates the long-term beneficial effects of trehalose in the protection of mouse sperm DNA by assuming that DNA degradation follows the first order rate kinetics. The current study will be helpful in optimization of mouse sperm desiccation solutions as well as storage conditions.

ACKNOWLEDGEMENTS
This work was supported by NIH grant no. 5U01HD38227-03.

REFERENCES

Bhowmick, S., Zhu, L., McGinnis, L., Lawitts, J., Nath, B. D., Toner, M., and Biggers, J., (2002), "Desiccation tolerance of murine spermatozoa dried at ambient temperature," *In submission to* Biology of Reproduction

Carpenter, J.F., Martin, B., Crowe, L. M., Crowe, J. H., (1987), "Stabilization of phosphofructokinase during air drying with sugar and sugar/transition metal mixtures," Cryobiology 1987; 24: 455-464.

Chen, T, Acker J. P, Eroglu, A., Cheley, S., Bayley, H., Fowler, A., Toner, M. (2001), " Beneficial effect of intracellular trehalose on the membrane integrity of dried mammalian cells," Cryobiology; 43: 168–181.

Crowe, J. H., Hoekstra, F. A., Crowe, L. M. (1992), 'Anhydrobiosis', Annul Rev Physiol 54: 579-599.

Crowe JH, Spargo BJ, Crowe LM., (1987), " Preservation of dry liposomes does not require retention of residual water," Proc. Natl. Acad. Sci. 1987; 84: pp. 1537-1540.

Evenson, D. and Jost, L., (2000), " Sperm chromatin structure assay is useful for fertility assessment," Methods in Cell Science, 22: 169-189.

Henriques Jr., F.C. (1947), " Studies of thermal injury, V. The predictability and the significance of thermally induced rate processes leading to irreversible epidermal injury," Arch. Path., 1947. 43: 489-502.

Kusakabe, H., Szczygiel, M.A., Whittingham, D.G., Yanagimachi, R. (2001), " Maintenance of genetic integrity in frozen and freeze-dried mouse spermatozoa," PNAS; 24: 13501-13506.

Puhlev, I., Guo, N., Brown, D. R., Levine, F. (2001), Desiccation tolerance in human cells. Cryobiology; 42: 207–217.

Wakayama T and Yanagimachi R. (1998), "Development of normal mice from oocytes injected with freeze-dried spermatozoa," Nature Biotech ; 16: 639-641.

Proceedings of IMECE2002
ASME International Mechanical Engineering Congress & Exposition
November 17–22, 2002, New Orleans, Louisiana

IMECE2002-33690

HEAT-INDUCED DENATURATION OF COLLAGENOUS TISSUE: A COMPARISON OF NUMERICAL SIMULATIONS WITH OCT AND MRI DATA

Alptekin Aksan[1*], Arthur F. Gmitro[2], Jennifer K. Barton[3], Scott Anseth[4], Steven P. Arnoczky[5], Damien J. Fron[1], John J. McGrath[1]

[1]Mechanical Engineering Department, Michigan State University, E. Lansing, MI 48824
[2]Department of Radiology, University of Arizona, Tucson, AZ 85724
[3]Division of Biomedical Engineering, University of Arizona, Tucson, AZ 85724
[4]Department of Orthopaedics, McLaren Regional Medical Center, Flint, MI 48532
[5]College of Veterinary Medicine, Michigan State University, E. Lansing, MI 48824
*Corresponding Author

ABSTRACT

Application of sub-ablative levels of heat to collagenous tissues has important therapeutic applications in medicine, such as tissue welding, thermokeratoplasty, skin resurfacing and treatment of joint instability. Sub-ablative heating produces collagen denaturation with desired tissue shrinkage yet detrimental comprises in mechanical properties of the tissue. In this paper, results of preliminary analyses with Optical Coherence Tomography (OCT) and Magnetic Resonance Imaging (MRI) are compared with a numerical model of tissue denaturation.

INTRODUCTION

A sub-ablative level of thermal damage produced in collagenous tissues is known to be the result of micro-scale structural transformations driven by the helix-coil transition of the collagen molecules (a review of the literature in this field can be found in [1]). Sub-ablative heat finds widespread use in tissue welding [2], thermokeratoplasty [3,4], skin resurfacing [5,6] and treatment of joint instability [7,8,9]. One such therapy of increasing interest is thermal capsulorrhaphy [10,11], during which lax collagenous tissues (the shoulder capsule and the underlying ligaments) surrounding the shoulder joint are shrunk by means of laser or radiofrequency heating [7,8,12]. Due to the differences in their modes of action, clinically utilized heating modalities (laser, monopolar and bipolar radiofrequency (RF)) create significantly different temperature distributions and thermal histories within the tissue. Development of a finite element analysis-based simulation tool to predict the sub-ablative thermal damage penetration in collagenous tissue was presented earlier [13]. In this paper, the thermal damage penetration predictions obtained by this simulation tool are compared to the results obtained from *in vitro* experimentation. Two different techniques were utilized to determine the extent of thermal damage into the tissue: Magnetic Resonance Imaging (MRI) and Optical Coherence Tomography (OCT).

MATERIALS AND METHODS

In order to verify the results of the numerical simulations, *in vitro* experimentation with equine flexor tendons was performed. A custom-made, computer-controlled stepper-motor driven carriage (Figure 1) was immersed in phosphate buffered saline (PBS) solution kept at 37°C by means of a constant temperature bath.

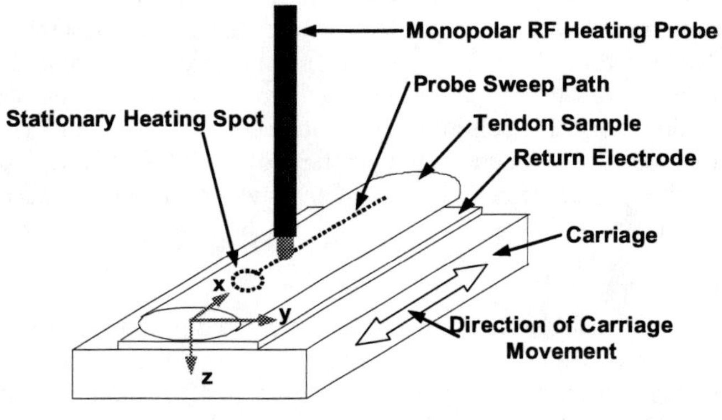

Figure 1: Experimental Setup for Monopolar RF Heating

In a typical experiment, a tendon sample (approximately, 70x30x5mm) was placed on the carriage and secured using two clamps on distal ends. A temperature-controlled monopolar RF heating probe (Vulcan TAC-S Electrothermal Probe, Oratec Interventions Inc., Menlo Park, CA) set to operate at a tip temperature of 75°C and a power setting of 25W was used to heat the tendon specimen. The return (ground) electrode was placed beneath the specimen (Figure 1). During the experiment the heating probe was kept stationary for 1 s and was then swept along the major axis (x-axis in Figure 1) of the tendon at a predetermined speed (0.5, 1, 2 or 4 mm/s). After the experiment, the sample was wrapped in PBS-soaked gauze and frozen below –80°C until Optical Coherence Tomography (OCT) and Magnetic Resonance Imaging (MRI) were performed. The MRI experiments were performed on a Bruker Biospec 4.7 T MRI instrument (Bruker BioSpin MRI Inc., Billerica, MA) equipped with a 32 mm RF coil using a 3D gradient echo sequence. The OCT system utilized a 1300 nm superluminescent diode with a 49 nm bandwidth, giving an axial resolution of 16 μm (details of the OCT system can be found in [14]). The lateral resolution of the system was set to approximately 13 μm by the imaging optics. Images were acquired at 14 A-scans (columns) per second. All images were taken at a lateral pixel size of 100 A-scans/mm and an axial pixel size of 512 points/A-scan. A-scans were kept constant at an optical depth of 3 mm. Assuming a tissue index of refraction of 1.4, the OCT signal was observed to disappear at approximately 2 mm in most specimens.

Details of the simulation method can be found elsewhere [2]. Briefly, by means of finite element (FE) analysis, the conservation of energy equation was solved for a three-dimensional control volume subjected to a moving heat source at its top boundary. The simulation algorithm accommodated the thermal effects of collagen denaturation as a solid-to-solid phase change. In order to simulate the tip-temperature controlled operation of the monopolar RF probe, the FE node temperature corresponding to the location of the probe (stationary or moving) was checked against the set temperature value (T_{set}=75°C) at each time increment. The heat source was turned off for the following time increment if the node temperature was within 2.5°C of the set value (hysteresis temperature). The hysteresis temperature is the threshold temperature difference value used to eliminate excessive hunting of the on/off temperature controller. (In the simulations, a hysteresis temperature of 2.5°C was assumed however, recent experiments show that the actual value is much higher).

RESULTS

The sub-ablative thermal damage created in the tissue is a function of temperature and exposure time. In the simulations, the thermal damage accumulated at each node in the FE mesh was calculated using the Arrhenius damage integral (Ω), as the thermal damage parameter. The simulation results for the thermal damage fields created are presented in Figures 2A and 3A, for probe sweep speeds of 2 mm/s and 4 mm/s, respectively. In the simulations, Ω=0.0 indicated no thermal damage and Ω=10.0 corresponded to complete thermal denaturation. In all of the figures presented, the black arrow indicates the location where the probe was kept stationary before it was swept and the gray arrow indicates the direction of the sweep.

Figure 2 shows the cross-sectional OCT image (A) and the MRI image (B) taken at the same location along the RF probe sweep path including the stationary heating spot. In both figures, the dashed lines indicate the detectable limit of thermal damage penetration. The effect of thermal denaturation, as observed in the OCT image is a change in the birefringence pattern of the tissue as indicated by the changing frequency of the dark/light patterns in the region surrounded by the dashed line (the damaged zone) and the outside (native tissue). The change in the birefringent pattern with thermal damage reflects a structural change, which is presumably the helix-coil transition of the collagen molecules. OCT predicted (Figure 2B) a thermal damage penetration of approximately 1.5mm at the stationary heating spot followed by a very shallow damage region (0.1-0.2mm) to the right along the sweep line.

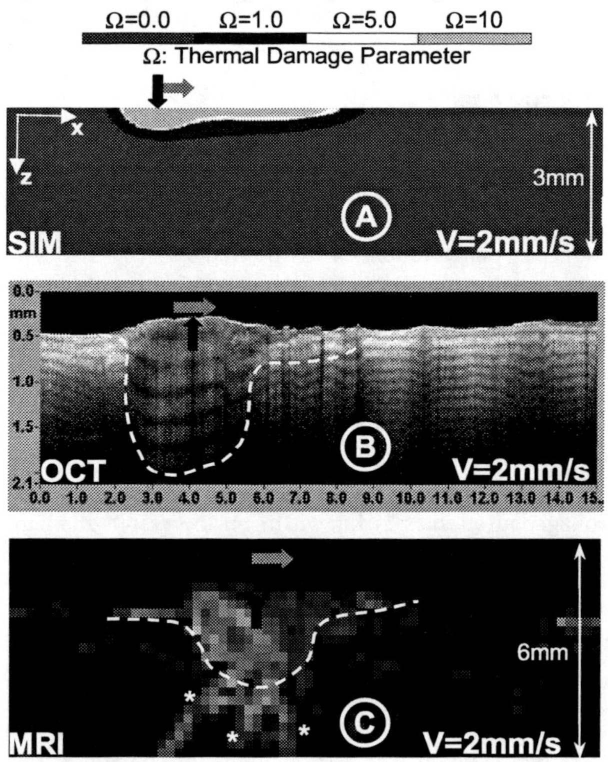

Figure 2: Comparison of the Simulation Prediction (A) with the OCT (B) and MRI (C) Images Showing the Depth of Thermal Damage (Ω) in the Tissue for V=2 mm/s.

The MRI image, owing to the difference in the relaxation times measured at the thermally damaged (appear as shades of gray) and native tissue regions (appear as black), predicted a uniformly damaged zone extending approximately 2.2 mm into the tissue. Below this zone, deeper into the tissue there appeared to be a finger-like pattern (indicated by *'s in Figure 2C). The reason for the existence of this finger-like pattern cannot be explained. The shape of the thermally damaged zone was predicted reasonably well with the simulation tool (Figure 2A) but the damage penetration depth was underestimated (the damage was approximately 0.75mm in the stationary heating spot and approximately 0.45mm along the sweep line.)

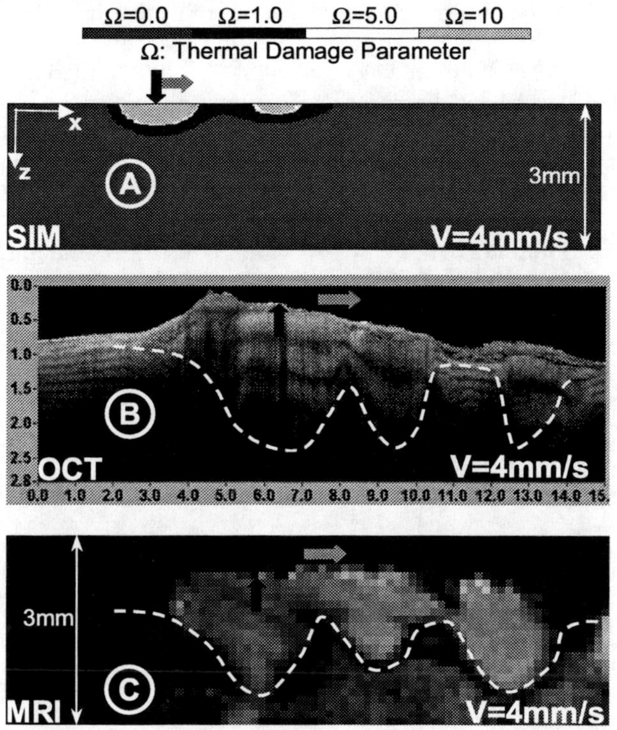

Figure 3: Comparison of the Simulation Prediction (A) with the OCT (B) and MRI (C) Images Showing the Depth of Thermal Damage (Ω) in the Tissue for V=2 mm/s.

In Figure 3A, the results are presented for the case of V=4mm/s. The shape of the thermal damage zone was distinctly different form the V=2mm/s case. Even though the thermal damage penetration at the stationary heating spot remained the same for both cases (as expected), along the sweep line a "peak and valley" trend is observed. Along this line the high damage zones were surrounded by lower damage regions. The same pattern was also observed in the simulation (only two of these regions are seen in Figure 3A since the simulation was stopped prior to the development of a third deeper zone). Depth of thermal damage decreased along the direction of probe motion and then increased further down the sweep path. In the OCT and MRI images (Figure 3B and 3C, respectively) three distinct high damage zones are seen. The first high damage zone corresponds to the stationary heating spot and the following two are along the sweep path. The existence of this interesting pattern may be explained by the interaction between the hysteresis temperature setting of the RF probe and the increase in the Peclet number with increased speed. In other words, the temperature at any fixed downstream location increases due to the heat conducted from the probe. Therefore, in the V=4mm/s case, the probe reaches the same downstream location twice as fast, the tissue temperature at that location (when the probe arrives) will be lower (when compared to the V=2mm/s case). This lower temperature may cause the controller to turn on and therefore may result in additional heating at that locale.

CONCLUSIONS

The preliminary results presented in this study have very important implications. Briefly, it was concluded that:

1. MRI and OCT have great potential to be applied as non-invasive tools to determine the extent of sub-ablative heat damage.
2. Even though the information that can be collected with OCT is limited to the superficial layers of the tissue (0-2.5mm), the high resolution obtained (16μm axial, 13 μm lateral) makes it unique to determine structural modification in superficial layers, such as those that would be created by lasers (e.g. during thermokeratoplasty and skin resurfacing). On the other hand, a spatial resolution of ~200μm is obtained with MRI. Another important factor is that OCT gives birefringence information, which is a direct indication of the tissue structure and is easily interpreted during heat induced collagen denaturation. However, more studies are needed to fine-tune this method to increase the penetration depth and to understand the effects of spatial nonuniformity of the optical properties of the tissue.
3. The simulation tool developed is very promising in determining the thermal damage penetration into the tissue and will be developed further to be used for three-dimensional therapy planning and heating probe design. The simulation tool underestimated the thermal damage, probably due to underestimation of the thermal conductivity of the tissue and overestimating the contribution of the convection heat transfer at the top surface. Additionally, recent experiments have now disclosed that the actual hysteresis temperature is higher than that assumed in the simulations. Future studies will focus on fine-tuning the numerical simulations through characterizing the thermal fields around the RF probes. Preliminary work aimed at this can be found in [15].

ACKNOWLEDGMENTS

The RF generator and the probes used in the study were lent to AA and JJM by Oratec Interventions Inc., Menlo Park, CA. The authors would like to thank Dr. Terri Zachos and Zachary Vaupel for their help during the experiments.

REFERENCES

[1] Arnoczky, S.P., Aksan, A., 2001, "Thermal Modification of Connective Tissues: Basic Science Considerations and Clinical Implications," *American Academy of Orthopaedic Surgeons Instructional Course Lectures*, FH Sims, F.H. (ed.), AAOS, Chicago, IL, **50**, pp.3-11.

[2] Cilesiz, I., Thomsen, S., Welch, A.J., 1997, "Controlled Temperature Tissue Fusion: Argon Laser Welding of Rat Intestine In Vivo, Part One," Lasers in Surgery and Medicine, **21**, pp. 269-277.

[3] Brinkmann, R., Radt, B., Flamm, C., Kampmeier, J., Koop, N., Birngruber, R., 2000, "Influence of Temperature and Time on Thermally Induced Forces in Corneal Collagen and The Effect on Laser Thermokeratoplasty," Journal Of Cataract and Refractive Surgery, **26**, 5, pp. 744-754.

[4] Kampmeier, J., Radt, B., Birngruber, R., Brinkmann, R., 2000, "Thermal and Biomechanical Parameters of Porcine Cornea," Cornea, **19**, 3, pp. 355-363.

[5] Kuo, T., Speyer, M.T., Ries, W.R., Reinisch, L., 1998, "Collagen Thermal Damage and Collagen Synthesis After Cutaneous Laser Resurfacing," Lasers in Surgery and Medicine, **23**, pp. 66-71.

[6] Fitzpatrick, R.E., Goldman M,.P., Satur, N.M., Tope, W.D., 1996, "Pulsed Carbon Dioxide Laser Resurfacing of Photoaged Facial Skin," Archives of Dermatology, **132**, pp. 395-402.

[7] Arnoczky, S.P., Aksan, A., 2001, " Thermal Modification of Connective Tissues: Basic Science Considerations and Clinical Implications," *American Academy of Orthopaedic Surgeons Instructional Course Lectures*, Sims, F.H. (ed.), AAOS, Chicago, IL, **50**, pp.3-11.

[8] Anderson, K., McCarty, E.C., and Warren, R.F., 1999, "Thermal Capsulorrhaphy: Where Are We Today," Sports Medicine and Arthroscopy Review, **7**, pp. 117-127.

[9] Perry, J.J., Higgins, L.D., 2000, "Anterior and Posterior Cruciate Ligament Rupture After Thermal Treatment," Arthroscopy, **16**, 7, pp. 732-736.

[10] Hecht, P., Hayashi, K., Lu, Y., Fanton, G.S., Thabit, G., Vanderby, R., Markel, M.D., 1999, "Monopolar Radiofrequency Energy Effects on Joint Capsular Tissue: Potential Treatment for Joint Instability," American Journal of Sports Medicine, **27**, **6**, pp. 761-771.

[11] Tyler, T.F., Calabrese G.J., Parker, R.D., Nicholas, S.J., 2000, "Electrothermally-Assisted Capsulorrhaphy (ETAC): A New Surgical Method for Glenohumeral Instability and Its Rehabilitation Considerations," Journal of Orthopaedic & Sports Physical Therapy, **30**, 7, pp.390-400.

[12] Tibone, J.E., McMahon, P.J., Shrader, T.A., Sandusky, M.D., Lee, T.Q., 1998, "Glenohumeral Joint Translation after Arthroscopic, Nonablative, Thermal Capsuloplasty with a Laser," American Journal of Sports Medicine, **26**, 4, pp. 495-498.

[13] Aksan, A., Nielubowicz, D.S., McGrath, J.J., 2001, "Modeling the Thermal Histories of Collagenous Tissues Subjected to Different Heating Modelities," HTD-21771, Proceedings of ASME 2001 International Mechanical Engineering Congress and Exposition.

[14] Hansen, K.A, Weiss, J.A., Barton, J.K., 2002, "Recruitment of Tendon Crimp with Applied Tensile Strain," Journal of Biomechanical Engineering, **124**, 72-77.

[15] Fron, D.J., Aksan, A., McGrath, J.J., 2002,"Ratiometric Laser-Induced Fluorescence (LIF) Thermography Of Radiofrequency Heated Tissue Phantoms, " HTD-21731, Proceedings of ASME 2002 International Mechanical Engineering Congress and Exposition.

Proceedings of IMECE2002
ASME International Mechanical Engineering Congress & Exposition
November 17–22, 2002, New Orleans, Louisiana

IMECE2002-33697

ESTIMATING PARAMETERS OF THERMAL DAMAGE FROM TRANSIENT EXPOSURES: LASER COAGULATION OF RAT LIVER

John A. Pearce
Dept. of Electrical and Computer Engineering
The University of Texas at Austin

Sharon Thomsen
Dept. of Biomedical Engineering
The University of Texas at Austin

ABSTRACT

Estimating effective thermal damage process coefficients for the first order kinetic model of damage processes is not difficult when the temperature is held constant for a substantial period. Laser coagulation experiments, however, are of short duration and, because of non uniform beam profiles, exhibit important heat transfer effects: the thermal histories are transient by nature and spatial temperature gradients are significant. An attempt is made to obtain useful estimates of activation energy, E, and collision frequency factor, A, directly from the transient history at the boundary of the zones of white coagulation and red hemorrhagic coagulation in liver in the rat, as identified by histologic studies. The Trial Region Method proves to be effective for transient data, in that a useful estimate of A and E can be obtained even from a single temperature curve. However, the results are very sensitive to small uncertainties in the actual value of Ω for the transient thermal histories to which the method is applied.

INTRODUCTION

First order Arrhenius kinetic models have been used to describe thermal damage processes since Henriques & Moritz classic studies of the 1940's [1]. The classical approach is to apply "constant" temperatures and to select experiments with "threshold damage" for analysis. The definition of threshold varies from study to study, and interpretation of the results is often somewhat subjective; however, the rate of heating is always an important consideration since the essential process is the denaturation of tissue proteins, rather than a phase change phenomenon.

NOMENCLATURE

- A Molecular collision frequency factor (s^{-1}).
- C Concentration of remaining native state tissue.
- E Denaturation activation energy (J/mole).
- R Universal gas constant = 8.3143 (J/mole/K).
- T Temperature (C or K).
- t Time axis (s).
- τ Total time of exposure (s)
- Ω Thermal damage parameter (no units).

METHODS

In classical formulations a dimensionless "damage parameter", Ω, is calculated from the reaction velocity expression:

$$\Omega(\tau) = \ln\left\{\frac{C(0)}{C(\tau)}\right\} = \int_0^{\tau} A\, e^{\left[\frac{-E}{RT}\right]} dt \qquad (1)$$

where the physical significance of Ω is the natural logarithm of the ratio of original concentration of undamaged protein, C(0), to the remaining undamaged protein at time, τ. The process parameters — collision frequency A and activation energy E — are determined experimentally. For constant temperature exposures the parameters A and E are determined by linear regression from a plot of ln{τ} vs 1/T (K^{-1}) over a range of exposure times for which Ω = 1, the classical threshold (i.e. ln{Ω} = 0). It is not always practical or possible to create essentially constant temperature exposures, particularly in the case of laser lesions. Transient thermal histories defy analysis by the classical, or any curve-fitting, approach to equation 1 since the root function is transcendental by nature.

The Trial Region Method

The Trial Region Method (TRM) evaluates a large number of combinations of A and E over a selected range against a "cost" for the predicted damage level (Ω), where the "cost" may be any parameter with ideal value = 0. The selected A and E range is searched for the minimum in the user-defined "cost" (*i.e.* the maximum in 1/cost). The method can be used on a collection of thermal histories or on a single transient temperature curve. An ideal successful combination of A and E yields Ω = 1 (ln{Ω} = 0) for a single T(t) record. In the case of an ensemble of transient curves with diverse heating times the (A, E) point which gives an intercept of ln{Ω} = 0, a slope of zero, and chi-squared = 0 on the regression line when the calculated ln{Ω} for each record is plotted *vs.* heating time, ln{τ}, is the best estimate for the process parameters. The user may select the relative weights to be given to each cost parameter (total cost = weighted combination of parameters).

RESULTS

Experiments

The TRM was applied to lesions in rat liver *in vivo* created by a Nd:YAG laser at 1064 nm [2]. The nominal laser spot size was 2 mm and powers from 3.2 to 10 W were applied for 5 or 10 s. Transient surface tissue temperatures were measured by infrared thermal camera and recorded on videotape for processing and digital calibration.

Transient records were obtained at the histologically-identified outer boundaries of thermal damage. The resulting lesions had three identifiable zones: 1) a central zone of severe thermal damage surrounded by 2) an annular hemispherical shell of white coagulation (thermal coagulation of structural proteins and blood), in turn surrounded by 3) an annular shell of red thermal damage (combined hyperemia, hemorrhage and hemostasis see Figure 1).

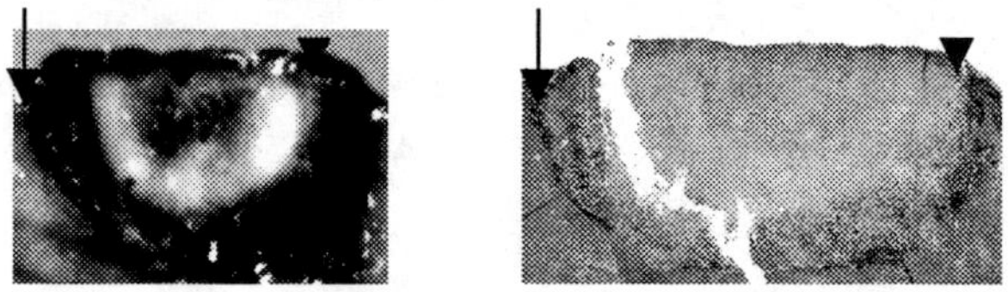

(a) (b)

Figure 1. Typical high energy lesion (expt 8-1). (a) Gross section, (b) H&E stained section; both showing the three damage regions. The surface Red/White zone boundary (arrowhead) is at 5.1 mm and the Red/Normal boundary (arrow) is at 7.1 mm diameter.

Maximum temperatures typical of the outer red zone/normal tissue boundary were 60 °C for 5 s, and 54 °C for 10 s laser activations.

TRM Convergence Behavior

The first problem is the establishment of "truth" for method calibration — primarily, the question of uniqueness for the solution point. The functional behavior of the TRM was studied on synthetic temperature curves which had known and specified values for A and E. The experimental temperature data closely fit a straight line when T is plotted *vs*. $\sqrt{t}$. The synthetic heating and cooling curves were constructed as two straight lines on $\sqrt{t}$ axes and integrated numerically using specified values of A and E. The slope of the heating curve was varied interactively until $\Omega = 1$ within a specified accuracy limit. The TRM was then applied to the synthetic transient data in the same fashion as for experimental data to determine whether the original A and E values could be recovered.

In the test sequence on single T(t) curves the TRM was consistently able to find a very sharp maximum in the (1/cost) plot for each curve — plotting (1/cost) in the ln{A}-E process plane hyperbolically enhances the convergence region making solution points easy to find. Typically, a combination of A and E with $|\ln\{\Omega\}| < 10^{-5}$ ($\Omega = 1.000010$ or 0.999990) was easily identified in these plots and is termed the "primary maximum". However, several discrete "secondary" (order of 10^{-3}) and even "tertiary" (order of 10^{-2}) and higher order maxima are also observed in a typical scan range of 10 for ln{A} (calculation at 151 points over the range) and 20 (kJ/mole) for E (at 201 points). Remarkably, all maxima, primary and higher order, fall on a single line in the process plane (E *vs* ln{A}) plot. The primary and other maxima stand out very clearly in a typical region scan: there is no ambiguity in the maximum location, adjacent values of A and E give much poorer "cost".

It must be kept in mind that if the search range includes a primary maximum, the TRM will find a match point to 10^{-5} in $\ln\{\Omega\}$ whether it is the actual A and E for the damage process or not. In other words, the input T(t) curve defines $\Omega = 1.0000$, and must be selected with extreme care. A hypothetical process with E = 70 kJ/mole and ln{A} = 24 was used to construct a heating curve 5 s long and subsequent cooling curve 2 s long, similar to the 5s data in the original experiment series. The slope of the heating $\sqrt{t}$ curve was varied interactively to obtain $\Omega = 0.999434$. However, this is not nearly close enough to 1.00000: the TRM found a primary and two secondary maxima for this T(t) curve with lower cost than the specified ln{A} and E values (see Table 1).

ln{A}	E (kJ/mole)	1/(|ln{Ω}|)	Maximum	Ω
21.33	62.78	15,730	Primary	1.000064
23.58	68.87	2,342	Secondary	1.000427
24	70	1,740	Tertiary	0.999425
26.25	76.09	977	Tertiary	0.998977
26.67	77.22	4,257	Secondary	1.000235

Table 1. TRM results for synthetic laser transient T(t) data: (1/cost) local maxima. Original process has only a tertiary maximum surrounded by lower order maxima on a common line.

The secondary and higher order maxima are separated by non uniform spacing on the line: ln{A} = 0.370 E (kJ/mole) – 1.89. Slight adjustment of the T(t) heating curve will make the primary maximum occur at (24, 70 kJ/mole), as intended.

The method works equally well on constant temperature data. For the same hypothetical process a 5s exposure at a constant temperature of 55.5949 °C results in $\Omega = 0.99999$. Table 2 summarizes the slightly different locus of maxima obtained: ln{A} = 0.366 E (kJ.mole) – 1.13.

ln{A}	E (kJ/mole)	1/(|ln{Ω}|)	Maximum	Ω
22.25	65.22	2,978	Secondary	0.99966
22.79	66.70	1,928	Tertiary	1.00052
23.46	68.52	1,147	Tertiary	0.99913
24	70	74,960	Primary	0.99999
25.21	73.30	1,839	Tertiary	0.99946
25.75	74.78	3,178	Secondary	1.00032

Table 2. TRM results for synthetic constant T data: (1/cost) local maxima. Original process has a clear primary maximum with other maxima on a common line.

Application of the TRM to Red/Normal Boundary Data

Given an arbitrary ensemble of T(t) curves it is not immediately clear how to best extract useful estimates for process parameters. One approach is to apply the TRM to the ensemble *en toto*: this assumes that each T(t) sample function represents $\Omega = 1$ precisely (or to at least equal precision). However, laser spots have steep radial thermal gradients and assignment of $\Omega = 1$ depends on simultaneous geometric calibration of histologic sections, thermographic images and surface laser spots — all on a total spatial scale of less than about 8 mm. There is additional uncertainty in the heating time, the inferred value of surface temperature, and the actual degree of damage in tissue. Positional uncertainty and degree of damage are subject to human interpretation. Further, damage boundaries occurring outside of the laser beam have thermal histories delayed with respect to the center of the beam on both the heating and cooling phases: temperature maxima often occur after the laser is turned off.

The TRM performs best when a single well-defined curve is used to estimate A and E. Attempts to find useful A and E values directly from cost functions based on the aggregate data did not yield satisfying results for this data set. The transient curves used in the calculations were obtained at the outer boundary of the red thermal damage zone. In the initial calculations the mean of $\ln\{\Omega\}$ and its standard deviation over the ensemble (both ideally zero) contributed equally to the calculated "cost"; however, an un-interpretable mish-mash zone of low cost was created by the interaction of multiple-order (1/cost) maxima over a fairly wide range on the $\ln\{A\}$-E convergence line. Consequently, the two exposure time values, 5s and 10s, were treated individually, and cost = $\ln\{\Omega\}$ for each activation time. Only two curves were available at 5s; these were averaged to obtain the best estimated T(t) record. Three experiments were averaged to describe the 10s activation.

The two 5 s curves each included a cooling time of 2 s, and the average curve used in the TRM calculation was also limited to a 2 s cooling time. No cooling curve interval was recorded for the 10s activations, so a 10s cooling curve with a $\sqrt{t}$ slope of -6.5 °C/$\sqrt{s}$ (matches the initial cooling segment, see Figure 2) was appended to facilitate the TRM calculation.

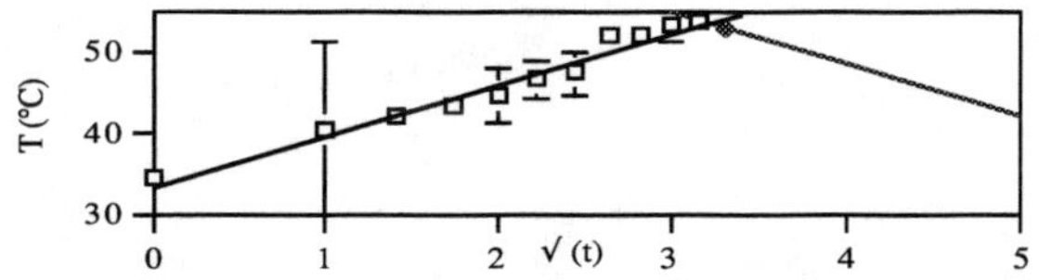

Figure 2. Heating and cooling curve segments for the 10s laser exposure. The heating curve is best fit by: $T = 33.4 + 6.361\sqrt{t}$ ($r^2 = 0.96$), the cooling curve by: $T = 74.5 - 6.479\sqrt{t}$. Error bars are ± Std. Dev.

A similar curve fit was applied to the 5s data. The heating curve best fit at 5s: $T = 20.597 + 16.811\sqrt{t}$ — note: this curve does not begin at t = 0 owing to the R/N boundary being outside of the laser beam; T = 32.4 °C at t = 0 for the heating curve. The cooling curve was extended from the original data (7s) out to 15 s using: $T = 93.447 - 15.323\sqrt{t}$.

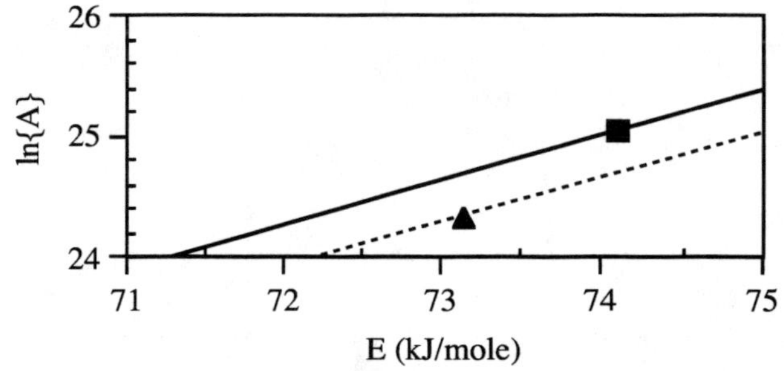

Figure 3. Estimated ln{A} and E from single-curve T(t) data at 5s (square) and 10s (triangle) durations. The respective locus of maxima lines are solid for the 5s calculation and dashed for the 10s calculation.

The TRM calculation estimated the ($\ln\{A\}$, E) points to be (25.04, 74.09) for the 5s data and (24.33, 73.13) for the 10s data, see Figure 3 for a plot. For model work one might reasonably use $\ln\{A\} = 24.69$ (i.e. $A = 52.5 \times 10^9$) and E = 73.6 kJ/mole, the average of these two values. The Trial Region Method has successfully rescued reasonable damage process parameters from relatively noisy data; as the difference between these two convergence points is quite small in damage process terms. The 5s (eq 2) and 10s (eq 3) locus of maxima lines are (respectively):

$$\ln\{A\} = 0.372\,E\ (\text{kJ/mole}) - 2.55 \qquad (2)$$

$$\ln\{A\} = 0.373\,E\ (\text{kJ/mole}) - 2.92 \qquad (3)$$

It is interesting that the locus of maxima lines are so close in slope and differ only slightly in offset. The physical significance of this is not clear at this point, but certainly warrants further investigation.

DISCUSSION

Transient thermal exposure experiments should be designed to provide the best possible estimates of T(t) curves for which $\Omega = 1.0000$. For example, multiple repeated experiments at one duration, power level and beam diameter are preferable to a wide distribution of durations power levels and beam sizes, in contra-distinction to the classical approach. One must scan several A and E regions on the locus of 1/cost maxima to be sure that the best (A, E) combination has been found. It is likely that several durations will give a better estimate of the centroid (A, E) point. However, collapsing the data to single curves, one representing each duration in the study, minimizes the overall uncertainties and provides the clearest, least ambiguous cluster of "best fit" damage parameter estimates. The TRM is equally applicable to data from any process described by equation 1, whether a thermal damage process or process of another kind. At least one T(t) curve for which $\Omega = 1.0000$ is all that is required.

The Trial Region Method gives astonishingly robust estimates of damage process parameters with surprisingly little input data for transient temperature exposures which cannot be studied by any other technique. However, its tendency for literal interpretation of the input data means that very certain records of T(t) and $\Omega = 1$ must be used to represent each duration studied. Signal averaging of input data prior to TRM calculation is preferable to collective analysis of all individual curves. Using the Trial Region Method we were able to extract useful damage process model parameters from an old study which was not originally designed, nor very well-suited, to provide them.

ACKNOWLEDGMENTS

We wish to thank John Bischof of the University of Minnesota for asking whether anything could be done to estimate damage process parameters from transient data. The thermographic data used in these calculations were calibrated by Tom McMurray and Brian Rae.

REFERENCES

1. Henriques, F.; Studies of thermal injury V: the predictability and significance of thermally induced rate processes leading to irreversible epidermal injury. *Arch. Pathol.* v43, pp489-502, 1947.
2. Thomsen, S., Schwartz, J., Joseph, R., Pearce, J., Rae, B., and McMurray, T.; Temperatures associated with thermally-induced red blood cells changes in tissues irradiated *in vivo*. *Proc. SPIE* v2130, pp156-163, 1994.

Proceedings of IMECE 2002
ASME International Mechanical Engineering Congress & Exposition
November 17-22, 2002, New Orleans, Louisiana

IMECE2002-33703

COMPARISON OF TEMPERATURE METRICS OF HEATING-INDUCED DAMAGE

Neil T. Wright

Department of Mechanical Engineering
Michigan State University
East Lansing, Michigan 48824-1226

INTRODUCTION

The temperature dependence of the rate of denaturation of Type I collagen due to heating is described well by chemical kinetics via the Arrhenius equation or transition state theory (TST) [1, 2, 3], each of which requires two material parameters. Nevertheless, many have sought to find a single convenient metric, such as one with units of temperature, to describe thermal denaturation of collagen. Comparing the results of studies that measured denaturation and cell death for a variety of biological samples shows that the parameters for either the Arrhenius equation (i.e. activation energy E_a and the frequency factor A) appear correlated over the range of temperatures for which biological materials are tested [4]. It has also been suggested that denaturation is a first-order phase change (i.e., melting) and thus should be characterized by a melting or denaturation temperature T_d [5].

Here, three metrics of thermal damage, each having units of temperature, are compared for collagen and collagenous tissues using available data. Each is defined in terms of parameters for TST, (i.e., activation enthalpy H_a and activation entropy S_a) or the Arrhenius equaton: the ratio of H_a to S_a to define T_d, the endothermic peak from differential scanning calorimetry (DSC) T_e, and a point on the damage integral T_c. How well each characterizes thermal stability of collagen is discussed.

DEFINITION OF METRICS

Burge and Hynes [6] suggest that Boedtker and Doty [7] were first to define T_d when Boedtker and Doty write of a "transition temperature" being the ratio of the "heat content change to the entropy change". Burge and Hynes thus write $T_d = H_a/S_a$. Others have defined T_d *ad hoc* based on measured quantities, such as T_d being the center of the temperature range for which shrinkage is quickest during a temperature scanning experiment [8]. During temperature scanning experiments, a sample is heated from some subphysiologic temperature to a supraphysiologic one, usually at a constant rate of temperature increase r $(= dT/dt)$. Typically, $1 \leq r \leq 10°\text{C min}^{-1}$ for $15 \leq T \leq 100°\text{C}$. (In temperature jump experiments, a specimen at a subphysiological temperature is heated quickly to a supraphysiological temperature and maintained there isothermally during the test.) Analyses of temperature scanning test results usually omit explicit time dependence of denaturation by transforming time to temperature via $T = T_0 + rt$ where T_0 is the temperature at the start of the test. Thus, changes with respect to time are written

$$\frac{d\Phi}{dt} = \frac{d\Phi}{dT}\frac{dT}{dt} = r\frac{d\Phi}{dT} \tag{1}$$

where Φ represents a dependent variable of interest. In DSC, the temperature at which the peak in the endotherm occurs T_e is usually specified as T_d. Miles [2] developed an implicit expression for T_e in terms of H_a, S_a, and r and assuming first-order kinetics, that is

$$\frac{1}{T_e} + \frac{E_a}{RT_e^2} = \frac{k_b T_e}{rh} e^{(S_a/R)} e^{(-H_a/RT_e)} \tag{2}$$

where k_b is the Boltzmann constant, h the Planck constant, and R the gas constant. Equation 2 indicates that T_e depends on r. Rate independent results are obtained as r is decreased [9].

Pearce and Thomsen [10] defined two other temperature-like indicators of thermal stability based on the Henriques description of the temperature dependence of denaturation [11]. The critical temperature T_c was defined as that temperature at which the time rate of change of the Henriques damage integral Ω is unity. That is

$$d\Omega/dt = Ae^{-E_a/RT_c} = 1 \tag{3}$$

Thus, $T_c = E_a/R\ln(A)$. Similarly, a threshold temperature T_t was defined as the temperature at which

$\Omega = 1$ for a given heating time. For isothermal heating, $T_t = E_a/R(\ln(tA))$, which equals T_c for a heating time of 1 s.

CALCULATION OF METRICS

Table 1 lists values of H_a and S_a taken from representative studies of the thermal denaturation of collagen and collagenous tissues. Also listed is the mean of the range of temperatures examined during each of these studies. All studies used temperature jump experiments except for Kampmeier [12], who used a DSC to scan the temperature from 30 to 95°C at 2 K/s. (Note that Miles [2, 3] used a DSC, but in a temperature jump mode when measuring the kinetics parameters.) The lowest temperature at which denaturation may be detected is often limited by experimental noise, while the upper one is limited by the time resolution of the data acquisition equipment or the heat transfer rate for spatial uniformity of the specimen. Thus, T_m could change with changed experimental design for these tests and is not a property of the specimens. Nevertheless, it provides a rough indication of the temperatures for which denaturation may be measured at experimentally relevant times.

Table 1. Values of H_a and S_a from several studies of the denaturation of collagen from various sources.

Specimen	T_m (°C)	H_a (kJ/mol)	S_a (kJ/mol K)	Ref.
tendon	60	586	1.450	[1]
lens capsule	53	858	2.380	[2]
tendon	59	518	1.485	[3]
swim bladder	27	836	2.760	[6]
swim bladder	28	339	0.961	[7]
cornea	65	106	0.039	[12]
tendon	55	367	0.823	[13]
tendon	78	354	0.754	[14]
joint capsule	66	231	0.355	[15]

Calculated values of T_d, T_c, and T_e appear in Table 2. For calculation of T_e, $r = (2/60)$ °C s^{-1} was used, consistent with the experiments of Kampmeier et al. [12]. Changing this to $r = (1/60)$ °C s^{-1} reduces the calculated values of T_e by less than 2°C for the samples listed, except for cornea, which drops from 49 to 43°C. This unusually large change results from the small value of S_a. The values of A and E_a (not shown) used to calculate T_c were calculated from $E_a = H_a + RT$ and $A = (k_b T_m/h)\ln(S_a/R)$.

DISCUSSION

Researchers have long sought a single parameter to characterize the thermal denaturation of collagen. Reported values of E_a and $\ln(A)$ for a variety of biological specimens are well correlated [4], suggesting that a single parameter could capture the temperature dependence of the rate of denaturation, even for an irreversible rate process. Conversely, Flory and Garrett [5], among others, have made thermodynamic arguments for collagen denaturation being a first-order phase transition and thus having a melting (i.e., denaturation) temperature. Others have shown that the same functional form for the temperature dependence of collagen denaturation results from considering denaturation to be an irreversible rate process [16]. Experimental design sometimes masks the explicit time dependence of denaturation, results then suggesting a single parameter such as T_d. For example Boedtker and Doty [7] measured an approximately 50% decrease in the intrinsic viscosity at about 28°C, with incipient change at 23°C and complete change at 33°C, each for 30 minutes heating. Other heating times were not reported. Although they calculate H_a and S_a, they conclude that there is "transition temperature" at about 30°C. In temperature scanning experiments, larger values of r can increase T_e [2, 9]. Even in temperature jump tests that consider changes in time explicitly, experimental design can obscure possible rate processes. Moran et al. [15], for example, note that above 63°C, their measurements of shrinkage are well characterized by first-order kinetics with $k(T)$ described by the Arrhenius relation. Below 63°C, measurement of shrinkage might be limited by experimental noise. Alternatively, they suggest that the stabilizing effects of other elements in the tissue samples (e.g., proteoglycans) may have decreased such that shrinkage occurs measurably at temperatures greater than 63°C.

Table 2. Comparison of temperature metrics.

Specimen	T_m (°C)	T_d (°C)	T_c (°C)	T_e (°C)	Ref.
tendon	60	131	71	65	[1]
lens capsule	53	88	54	50	[2]
tendon	59	76	66	21	[3]
swim bladder	27	30	5.4	3	[6]
swim bladder	28	79	5.9	0	[7]
cornea	65	2450	89	48	[12]
tendon	55	174	71	59	[13]
tendon	78	197	78	69	[14]
joint capsule	66	378	106	87	[15]

The metrics compared here are each defined in terms of rate parameters and have units of temperature. Each roughly follows the trend of T_m, although T_d has several points that are unrealistically high. Conversely, T_c and T_e each follow T_m closely, suggesting that these reflect better the thermal stability of collagen; T_c is somewhat better correlated with T_m. Both T_c and T_e are defined as specific points in a first-order rate process that describes well the thermal denaturation of collagen. An additional assumption about r is needed to calculate T_e. Thus, T_c, based on an arbitrary but fixed level of the damage rate, appears to be the best of these three metrics at representing the thermal stability of collagen.

REFERENCES

[1] Weir, C.E., 1949, "Rate of Shrinkage of Tendon Collagen - Heat, Entropy, and Free Energy of Activation of the Shrinkage of Untreated Tendon. Effect of Acid, Salt, Pickle, and Tannage on the Activation of Tendon Collagen ", *J. Am. Leather Chem. Assoc.*, **44**, pp.108-140.

[2] Miles, C.A., 1993, "Kinetics of Collagen Denaturation in mammalian lens capsules studied by differential scanning calorimetry", *Int. J. Biological Macromolecules*, **15**, pp. 265-271.

[3] Miles, C.A., Burjanadze, T.V., and Bailey, A.J., 1995, "The Kinetics of the Thermal Denaturation of collagen in unrestrained rat tail tendon determined by differential scanning calorimetry", *J. Molecular Bio.*, **245**, pp. 437-446.

[4] Wright, NT, 2002, "On a Relationship Between the Arrhenius Parameters from Thermal Damage Studies", *Journal of Biomechanical Engineering*, accepted.

[5] Flory, P.J. and Garrett, R.R., 1958, "Phase Transitions in Collagen and Gelatin Systems", *J. Am. Chem. Soc.*, **80**, pp. 4836-4845.

[6] Burge, RE and Hynes, RD, 1959, "The Thermal Denaturation of Collagen in Solution and its Structural Implications", *J. Molecular Bio.*, **1**, 155-164.

[7] Boedtker, H and Doty, P, 1956, "The Native and Denatured State of Soluble Collagen", *J. Am. Chem. Soc.*, **78**, pp. 4267-4280.

[8] Gustavison, KH, 1956, *The Chemistry and Reactivity of Collagen*, Academic Press, New York.

[9] Lepock, JR, Ritchie, KP, Kolios, MC, Rodahl, AM, Heinze, KA, and Kruub, J, 1992, "Influence of Transition Rates and Scan Rate on Kinetic Simulations of Differential Scanning Calorimetry Profiles of Reversible and Irreversible Protein Denaturation", *Biochemistry*, **311**, pp. 12706-12712.

[10] Pearce, J. and Thomsen, S., 1995, "Rate Process Analysis of Thermal Damage," in: *Optical-Thermal Response of Laser-Irradiated Tissue*, A.J. Welch and M.J.C. van Gemert, eds., Plenum Press, New York.

[11] Henriques, F.C., Jr., 1947, "Studies of Thermal Injury V.: The predictability and the significance of thermally induced rate processes leading to irreversible Epidermal Injury", *Archives of Pathology*, **43**, pp. 489-502.

[12] Kampmeier, J, Radt, B, Birngruber, R., Brinkmann, R., 2000, "Thermal and Biomechanical Parameters of Porcine Cornea", *Cornea*, **19**, pp. 355-363.

[13] Maitland, D.J., and Walsh, J.T., 1997, "Quantitative Measurements of Linear Birefringence During Heating of Native Collagen", *Lasers Surgery Medicine*, **20**, pp. 310-318.

[14] Chen, S.S., Wright, N.T., and Humphrey, J.D., 1997, "Heat-induced Changes in the Mechanics of a Collagenous Tissue: Free Shrinkage", ASME *J. Biomech. Engr.*, **119**, pp. 372-378.

[15] Moran, K., Anderson, P, Hutcheson, J., Flock, S., 2000, "Thermally Induced Shrinkage of Joint Capsule", *Clinical Orthopaedics and Related Research*, **381**, pp. 248-255.

[16] Miles, CA and Ghelashvili, M, 1999, "Polymer-in-a-Box Mechanism for the Thermal Stabilization of Collagen Molecules in Fibers", *Biophysical Journal*, **76**, pp. 3243-3252.

Proceedings of IMECE2002
ASME International Mechanical Engineering Congress & Exposition
November 17–22, 2002, New Orleans, Louisiana

IMECE2002-32044

EFFECT OF PHYSIOLOGICAL PARAMETERS ON THE TEMPERATURE DISTRIBUTION OF A LAYERED HEAD MODEL

Obdulia Ley

Department of Mechanical Engineering
Rice University
Houston, Tx 77005-1892
Email: obdulia@ruf.rice.edu

Yildiz Bayazitoglu

Department of Mechanical Engineering
Rice University
Houston, Tx 77005-1892
Email: bayaz@ruf.rice.edu

ABSTRACT

Brain temperature control is important in clinical therapy, because moderate temperature reduction of brain temperature increases the survival rate after head trauma. A factor that affects the brain temperature distribution is the cerebral blood flow, which is controlled by autoregulatory mechanisms.

To improve the existing thermal models of brain, we incorporate the effect of the temperature over the metabolic heat generation, and the regulatory processes that control the cerebral blood perfusion and depend on physiological parameters like, the mean arterial blood pressure, the partial pressure of oxygen, the partial pressure of carbon dioxide, and the cerebral metabolic rate of oxygen consumption.

The introduction of these parameters in a thermal model gives information about how specific conditions, such as brain edema, hypoxia, hypercapnia, or hypotension, affect the temperature distribution within the brain.

Existing biological thermal models of the human brain, assume constant blood perfusion, and neglect metabolic heat generation or consider it constant, which is a valid assumption for healthy tissue. But during sickness, trauma or under the effect of drugs like anesthetics, the metabolic activity and organ blood flow vary considerably, and such variations must be accounted for in order to achieve accurate thermal modeling.

Our work, on a layered head model, shows that variations of the physiological parameters have profound effect on the temperature gradients within the head.

INTRODUCTION

In this work, we study the effect of the temperature and the physiological parameters on the cerebral blood perfusion W_b, and introduce the relation between the temperature and the tissue metabolic activity q_m. We present the changes in the temperature distribution for various values of the mean arterial blood pressure ($MABP$), the partial pressure of oxygen (P_{O_2}), the partial pressure of carbon dioxide (P_{CO_2}), and the cerebral metabolic rate of oxygen consumption ($CMRO_2$). Variations of these parameters are used by clinicians during surgery, like cadiopulmonary bypass (CPB), or to treat patients with acute brain damage.

Tissue temperature distribution is affected by blood perfusion. In the brain, the cerebral blood flow (CBF) is controlled by regulatory mechanisms, which depend on physiological parameters like the $MABP$, Pa_{O_2} and Pa_{CO_2}.

Adequate cerebral blood flow is essential to maintain organ function (Busija, 1993); therefore, to asses therapy and prevent brain injury after trauma, in sickness or during surgery, clinicians continuously monitor these physiological parameters.

Several laboratory animals have been used to study brain cooling, and the relation between brain temperature and energy utilization. Brain temperature gives information about the extent of neuronal injury after stroke (Laptook, 1997), and moderate temperature reduction of brain tissue decreases tissue damage after ischemia. However, the optimal method to reduce brain temperature is unknown (Laptook, 2001), and direct measurement of brain temperature is highly destructive and can be performed only in a few surgical cases. Therefore, developing accurate thermal models is necessary to improve clinical therapies involving temperature.

The introduction of physiological parameters in a thermal model gives information about how specific conditions, produced by trauma, drugs or sicknesses, affect the temperature distribution within the brain. To our knowledge, the effect of the regulatory mechanisms and physiological parameters on the cerebral blood perfusion, has not been studied before.

The consideration of physiological parameters in a thermal model is important because it will allow us to create more accurate thermal models, and will help clinicians to incorporate temperature to the control and diagnostic techniques used in the treatment of head trauma.

Mathematical Model

To determine the steady state temperature distribution, the head is modeled as a layered hemisphere. We use a Pennes like bioheat equation (Pennes, 1948),

$$k_i \frac{d^2T_i}{dr^2} + \frac{2}{r}\frac{dT_i}{dr} + \rho_b c_b W_b (T_a - T_i) + q_m = 0, \tag{1}$$

where $i = 1,2,...,M$, and M is the number of layers, W_b represents the blood perfusion term, and q_m denotes the metabolic heat generated by the tissue. We assume a constant temperature T_o at the external surface, symmetry about the origin and that all the properties are isotropic within each layer. Finally, we assume continuity of temperature and heat flux on the layer interfaces located at $r = r_i$, for $i = 1,2,...,M-1$.

To improve the available thermal models (Xu, 1999)-(Xu, 1999), we consider that W_b and q_m are functions of the temperature, position and the physiological parameters $\phi_i = PaCO_2, PaO_2$, and CMRO_2. This dependence is introduced in our thermal model based on the clinical observations that changes in the concentration of CO_2 in the brain tissue affect the time required to achieve brain cooling during hypothermic treatment (Conn, 1995)-(Payne, 1963).

Tissue Metabolism and Physiology

Blood circulation through a tissue allows the transport of substances such as oxygen and glucose, products of the metabolic activity, and hormones. In the brain, there exists regulatory mechanisms that help maintain adequate blood flow by dilating or constricting the diameter of the small arteries depending on the concentrations of the physiological parameters.

Brain autoregulation helps keep the CBF constant despite changes in the MABP, and also maintains the CBF depending on the level of activity within the tissue.

Efficacy in cerebral protection during cardiothoracic surgery or after head trauma depends largely upon metabolic suppression and control of cerebral blood flow. For these reasons, clinicians continuously measure the CMRO_2, $PaCO_2$, and PaO_2.

The metabolic heat released by a tissue q_m is proportional to the tissue oxygen consumption. In humans, the cerebral metabolic rate of oxygen consumption (CMRO_2) increases exponentially with the temperature (Siesjo, 1978), and linearly with the CBF for temperatures lower than the protein break down temperature (42^oC). Thus, the tissue metabolic activity can be expressed as

$$q_m(T) = q_m^o \exp(c\Delta T), \tag{2}$$

where q_m^o is the tissue metabolic heat generated at normal core

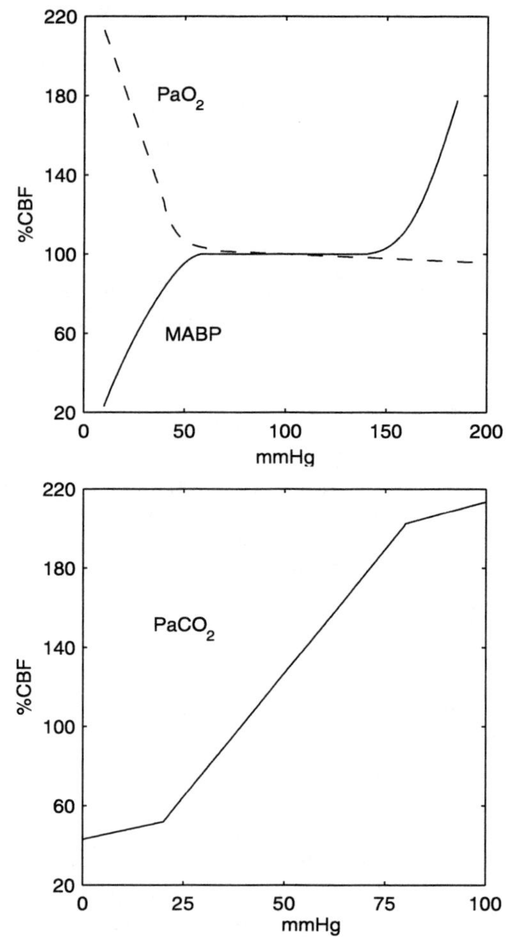

Figure 1. Variations of the CBF with MABP, $PaCO_2$, and PaO_2. The normal average values for these parameters are: MABP=100 mmHg, PaO_2=100 mmHg, and $PaCO_2$=40 mmHg.

temperature; and its values for the different tissue types are reported in the literature (Zhu, 2001); ΔT represents the temperature difference between the tissue temperature T and the normal core temperature, or the arterial temperature T_a, *i.e.* $\Delta T = T - T_a$. The constant c depends on the metabolic activity of the tissue, it varies depending on the tissue type, and is obtained by interpolating tissue oxygen consumption curves and using the Q_{10} law (Siesjo, 1978).

Studies using hypothermic cadiopulmonary bypass and direct blood cooling show that, an increment in the Pa_{CO_2} produces a more rapid temperature fall in the brain temperature (Conn, 1995)-(Payne, 1963). This is due to the reduction in the cerebrovascular resistance produced by the high CO_2 tension.

Because of the relation between physiological parameters and blood flow, and the importance of the blood flow in the temperature of a perfused tissue, we incorporate the regulatory mechanisms in our thermal model of the head.

We introduce the autoregulation mechanism in the thermal model by defining the cerebral tissue perfusion W_b as a function of the temperature, position and the systemic parameters (PaO_2, $PaCO_2$, and CMR_{O_2})

The relations between the cerebral blood perfusion W_b and the physiological parameters are obtained from a cerebral regulation model (Thoman, 1998), and are incorporated as follows

$$W_b(\phi, T, r) = W_o\left(1 + \Delta CBF_\phi\right), \tag{3}$$

where W_o is the average blood perfusion in the cerebral tissue at normal core temperature; ΔCBF_ϕ represents the percentile change in the cerebral blood flow due to variations in the physiological parameter ϕ_i, and it is defined as

$$\Delta CBF_\phi = \sum_{\phi_i}\left(\frac{\%CBF_{\phi_i} - 100}{100}\right), \tag{4}$$

for $\phi_i = MABP$, Pa_{O_2}, Pa_{CO_2}, and $CMRO_2(T)$). The expressions for the functions $\%CBF_{\phi_i}$ are documented in the literature (Thoman, 1998), and plotted in Fig. 1.

The exponential model for $q(T)$ was known by physiologists since 1944 (Siesjo, 1978); however, it has been considered only recently by Zhu *et al* (Zhu, 2001) in a thermal model of the head. In this work, we include not only the temperature dependent metabolic heat described in equation 2, but we introduce for the first time the effect that some physiological parameters have on the organ blood flow and the tissue temperature distribution. This is last is done by using the available organ regulatory models.

Table 1. Physical and physiological parameters for the 3-layer head model

	Brain	Skull	Skin
k_i (W/m^oC)	0.50	1.16	0.34
ρ_i (kg/m^3)	1050	1500	1000
c_i (J/kg^oC)	3700	2300	4000
W_o (ml/min 100 grs of tissue)	50.02	0.1	2.0
q_o (W/m^3)	10437	368.3	363.4
Thickness (mm)	85	4	4

For the blood, $c_b = 3800\ J/Kg^oC$, and $\rho_b = 1050\ Kg/m^3$

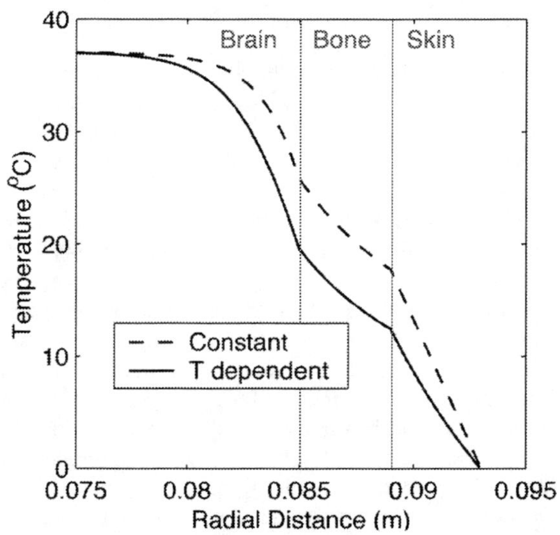

Figure 2. Temperature distribution for a 3-layer hemisphere (skin/skull/brain) with arterial temperature $Ta = 37^oC$, and external skin temperature $T_o = 0^oC$. Dashed line: $W_b = W_o, q_m = q_o$ and Solid line: $W_b = W(T), q_m = q_m(T)$.

Results

The temperature dependence of the metabolic heat and blood perfusion, gives a nonlinear, second-order differential equation which is solved numerically using the Gauss-Seidel iterative method.

We calculated the temperature distribution for a 3-layer hemisphere formed by skin, skull and brain tissue((Zhu, 2001)-(Xu, 1999)), subject to a constant temperature boundary condition at the skin surface ($T_o = 0$). In our calculations, the metabolic heat generation q_m and the tissue blood perfusion W_b

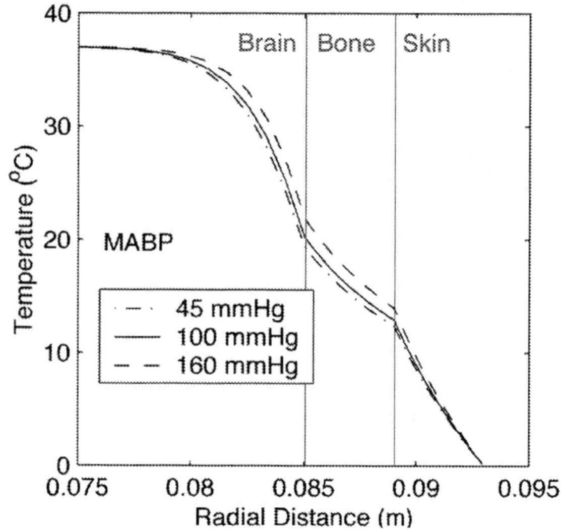

Figure 3. Temperature distribution for a 3-layer hemisphere (skin/skull/brain) with $Ta = 37^oC$, $T_o = 0^oC$. $PaCO_2$= 40 mmHg, PaO_2= 100 mmHg, and *MABP*= 45, 100, 160 mmHg.

for the brain tissue region are given by equations 2 and 3, respectively. In the skin and skull layers, q_m and W_b are assumed constant, and their values are given in Table 1.

To observe the effect of the temperature dependent properties on the head temperature distribution, we consider the constant and the temperature dependent metabolic heat in the brain. The temperature distribution for each case is shown in Fig. 2. In Fig. 2, the dashed line denotes constant blood perfusion and metabolic heat, and the solid line shows the temperature distribution for q_m and W_b given by equations 2 and 3. For this calculation, normal average values of the physiological parameters were considered.

From Fig. 2, one observes that the temperature distribution in the deep structures of the brain (gray and white matter) varies only within a small distance d from the interface between brain and skull. This distance d wil be referred as *penetration depth*. d is small due to the high blood perfusion of the brain tissue, and increases about 15 % when the temperature dependence of W_b and q_m is considered. Also, the temperature distribution in the outer structures (skin and skull) for the temperature dependent W_b and q_m, is lower than the constant W_b and q_m case, by as much as 3^oC in every point, and this temperature difference decreases in deeper structures where both temperature distributions converge to the a limiting temperature very close to the arterial temperature T_a. This is due to the large volume of warm arterial blood continuously perfusing the white and gray matter. Thus, if the temperature of the deep tissue is to be altered, the arterial blood temperature must be changed. Variation of the arterial blood temperature is usually done through extracoporeal perfusion.

In Fig. 3, we show how variations of the MABP affect the temperature distribution. Variations of the *MABP* is produced by drugs, or conditions like brain edema, brain swelling, vessel collapse or aneurysms. We consider values of the MABP that correspond to mild hypothension (MABP= 45 mmHg), normotension (MABP=100 mmHg), and mild hypertension (MABP= 160 mmHg), and maintain the rest of the physiological parameters in their average values. In the low pressure case the blood flow decreases 10% approximately producing a light temperature drop . When the MABP=160 mmHg, the CBF increases 13%, and the temperature in in the bone layer and the outer part of the brain increases up to 2^oC. The penetration depth d varies between 3% to 9% in each case.

In Figs. 4 and 5, we show the variation of the radial temperature distribution for different values of Pa_{O_2} and Pa_{CO_2}, keeping the other physiological parameters in their average values and using the exponential temperature dependence (3) for the $CMRO_2$ and the metabolic heat q_m. Variations of Pa_{O_2} and Pa_{CO_2} are achieved by respiration of air with different concentrations of either O_2 or CO_2.

In Fig. 4 we consider values of Pa_{O_2}= 25 and 50 mmHg, which produce a CBF increase of the 70% and 7%, respectively; while Pa_{O_2}= 250 mmHg reduces the CBF only 2% of its average value (Pa_{O_2}= 100 mmHg). It is observed that the Pa_{O_2} variation in the range 25-250 mmHg produces a maximum local temperature change of about 15% and this change occurs at the interface between bone and brain, also, the penetration depth d is reduced up to 35 % when Pa_{O_2}=25 mmHg.

In Fig. 5, the CBF in increased form 100% at Pa_{CO_2}=40 mmHg, to 208% when Pa_{CO_2}= 90 mmHg. In this case the maximum local temperature change is less than 18%, and occurs at the same location as before. The penetration depth d is reduced the order of 43% when Pa_{CO_2}=90 mmHg. Pa_{CO_2} is a strong vasodilator of the cerebral vasculature, and it is able to increase the CBF by 100% during severe hypercapnia. Hypocapnia, on the other hand, reduces CBF and improves the cerebral autoregulatory capacity.

Any change in the cerebral blood volume, produced as a result of the regulatory mechanisms, will affect the head temperature distribution. it is observed that varying the different physiological parameters so that the CBF is increased, the temperature gradient in the external structures and the penetration depth d are reduced. However, the ability to control the deep tissue temperature of a patient will strongly depend on the the temperature of the arterial blood entering the brain, and the external boundary conditions applied to the head and neck surface.

The results of this model can be used by anesthesiologists, pediatricians, radiologists, and in the development of teaching models that will help medical students to evaluate the severity of a condition and to search for optimal treatment incorporating

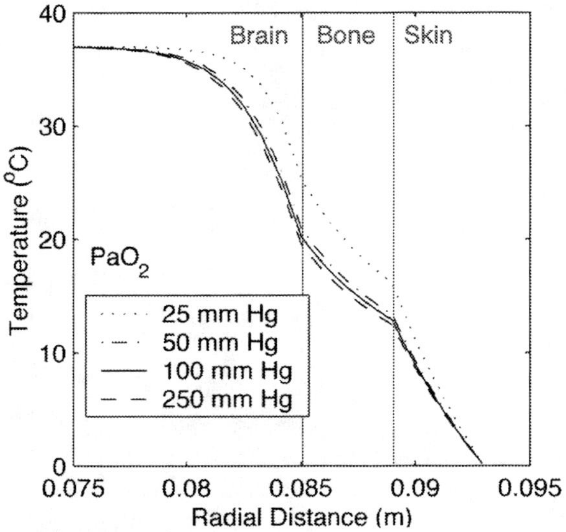

Figure 4. Temperature distribution for a 3-layer hemisphere (skin/skull/brain) with $Ta = 37^oC$, $T_o = 0^oC$, and normal MABP=100 mmHg, for a) PaO_2= 100 mmHg, and $PaCO_2$= 40, 60 and 90 mmHg

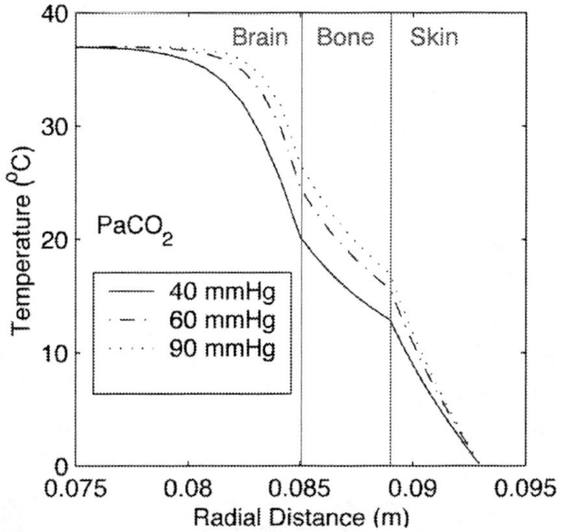

Figure 5. Temperature distribution for a 3-layer hemisphere (skin/skull/brain) with $Ta = 37^oC$, $T_o = 0^oC$, and normal MABP=100 mmHg, for $PaCO_2$= 40 mmHg and PaO_2= 25, 50, 100, 250 mmHg.

temperature as one variable.

During brain injury or other pathological conditions, instabilities of the intracranial pressure, cerebral blood volume and CBF occur. The time variation of these quantities is obtained from intracranial dynamics models (Ursino, 1997), wich will be implemented to the time dependent bioheat equation as future work.

ACKNOWLEDGMENT

The authors would like to thank Dr. Matt Bayazitoglu for his comments and guidance throughout this work.

REFERENCES

D.W. Busija, *Cerebral Autoregulation*, in *The Regulation of Cerebral blood Flow*, Edited by J. Willis, CRC Press, pp 45-60, 1993.

A.R. Laptook, R.J. Corbett, R. Sterret, D.K. Burns, Garcia, D. and G. Tollesfbol, *Modest Hypothermia Provides Partial Neuroprotection When Used for Immediate Resuscitation after Brain Ischemia*, Pediatric Research, Vol. 42, No. 1, pp. 17-23, 1997.

A.R. Laptook, L. Shalak, and R.J. Corbett, *Differences in Brain Temperature and Cerebral Blood Flow During Selective Head versus Whole-Body Cooling*. Pediatrics, Vol. 108, No. 5, pp. 1103-1110, 2001.

B.K. Siesjo, *Brain Energy Metabolism*, John Wiley and Sons, New York, 1978.

L. Zhu and C. Diao, *Theoretical Simulation of Temperature Distribution in the Brain during Mild Hypothermia Treatment for Brain Injury*, Proceedings of the ASME, 35th National Heat transfer Conference, Anaheim, California, 2001.

X. Xu, P. Tikuisis, and G. Giesbrecht, *A Mathematical Model for Human Brain Cooling during Cold-Water Near-Drowning*, Journal of Applied Physiology, Vol. 86, No. 1, pp. 265-272, 1999.

W.J. Thoman, S. Lampotang, D. Gravenstein, and J. Aa, *A Computer Model for Intracranial Dynamics Integrated to a Full-Scale Patient Simulator*, Computers and Biomedical Research, Vol. 31, pp. 32-46, 1998.

H. Pennes, *Analysis of Tissue and Arterial Blood Temperature in the Resting Human Forearm*, Journal of Applied Physiology, Vol. 1, No. 2, 1948, pp. 93-125.

A.W., Conn, K. Miyakasa, M. Katayama, M. Fujita, H. Orima, G. Barker, and D. Bhon, *A Canine Study of Cold Water Drowning in fresh versus salt water*, Critical Care Medicine, Vol. 23, No. 12, pp 2029-2037, 1995.

W.S., Payne, R.A., Theye, and J.W., Kirkland, *Effect of CO_2 on the Rate of Brain Cooling during the Introduction of Hypothermia by Direct Blood Cooling*, Journal Of Surgical Research, Vol. 3, pp. 54-60, 1963.

M. Ursino, and C.A. Lodi, *A Simple Mathematical Model of the Interaction between Intracranial Pressure and Cerebral Hemodynamics*, Journal of Applied Physiology, Vol. 82, pp. 1256-1269, 1997

Proceedings of IMECE2002
ASME International Mechanical Engineering Congress & Exposition
November 17–22, 2002, New Orleans, Louisiana

IMECE2002-32045

FINITE ELEMENT SIMULATION OF COOLING OF REALISTIC 3-D HUMAN HEAD AND NECK

Brian H. Dennis [1]
Institute of Environmental Studies
Graduate School of Frontier Sciences
University of Tokyo
7-3-1 Hongo, Bunkyo-ku, Tokyo 113-8656, JAPAN
Tel: +81-3-5841-8922 ; FAX: +81-3-5841-8922
dennis@garlic.q.t.u-tokyo.ac.jp

Robert C. Eberhart [2]
Department of Biomedical Engineering
500 West First Street
The University of Texas at Arlington
Arlington, Texas 76019, U.S.A.
Tel: (214) 648-2052; FAX (214) 648-2979
Robert.Eberhart@UTSouthwestern.edu

George S. Dulikravich [3]
Department of Mechanical and Aerospace Engineering
Multidisciplinary Analysis, Inverse Design & Optimization
The University of Texas at Arlington, UTA Box 19018
Arlington, Texas 76019, U.S.A.
Tel: (817) 272-7376; FAX: (817) 272-5010
gsd@mae.uta.edu

Steve W. Radons [4]
Research & Development
Medtronic Physio Controls Corporation
11811 Willows Road NE, P. O. Box 97006
Redmond, Washington 98073-9706, U.S.A.
Tel: (425) 867-4427; FAX: (425) 867-4462
stephen.radons@physio-control.com

ABSTRACT

Rapid cooling of the brain in the first minutes following the onset of cerebral ischemia is a potentially attractive preservation method. This computer modeling study was undertaken to examine brain-cooling profiles in response to various external cooling methods and protocols, in order to guide the development of clinical cooling devices. The criterion of successful cooling is the attainment of a 33.0 °C average brain temperature within 30 minutes of treatment. Comparison of the finite element model results with a formal mathematical solution, give confidence that the simulation methods are sound.

The cooling simulations considered to date all indicate that no one means of external cooling of the head or neck is sufficient to cool the brain in a reasonable period of time (30 minutes).

Neither ice packs applied to head or neck, or cooling helmets can satisfy the 33.0 °C target temperature specification.

This central conclusion of insubstantial cooling is supported by the modest enhancements reported in experimental investigations of externally applied cooling.

The key problem is overcoming the protective effect of warm blood perfusion, which reaches the brain via the uncooled carotid arterial supply and effectively blocks the external cooling wave from advancing to the core of the brain. This suggests that other cooling means should be explored requiring a realistic simulation of cooling of other pertinent parts of the human anatomy.

INTRODUCTION

This computer modeling study was undertaken to examine brain-cooling profiles in response to various external cooling methods and protocols. The overall goal is to guide the development of devices to obtain a 33.0 °C average target temperature in the brain within 30 minutes of treatment [1].

The reason for cooling the brain is described by a well-known "Q_{10} law" of thermal physiology. This law states that for every ten degrees Celsius reduction in tissue temperature there is a corresponding reduction in cell metabolism equal to Q_{10}. Values for Q_{10} range from 2.0 to 3.0 and are cited in the physiology literature. Formally, the law is written as

$$\frac{q_m}{q_{m0}} = Q_{10}\, exp[(\theta - \theta_o)/10] \qquad (1)$$

Assuming a mid range value, say 2.5, it is seen that reducing tissue temperature from 37.0 °C to 27.0 °C reduces

[1]Research Associate. Member ASME.
[2]Professor of Engineering in Surgery. Fellow ASME.
[3]Professor. Director of MAIDO Program. Fellow ASME.
[4]Manager.

the metabolic rate by a factor of about 2.7. These values underscore the promise of cooling for cerebral protection.

We used data from the National Library of Medicine model *Standard Man* [2], derived from 1 mm thick sections of the entire cadaver of an executed male convict. A three-dimensional tetrahedral mesh of the head was generated from these data by researchers at the University of Karlsruhe [3,4]. It was composed of 43448 linear tetrahedrons with 14 material property domains, including air cavities within the head. The triangular surface mesh is shown in Figure 1. The underlying voxel representation we used is the result of the tissue classification of the Visible Man Dataset by the Karlsruhe University team [3,4]. Capillary "tissue" perfusion data from an experimental study of rapid brain cooling in a subhuman primate model at the University of Texas Southwestern Medical Center at Dallas [5,6] was adapted to human tissue types.

Consensus values of thermal parameters for various body tissues were used, as given in Table 1 [7,8,9,10]. The steady state and transient Pennes equations were discretized among all computational nodes and solved for various initial, boundary and perfusion conditions.

In addition, an idealized neck problem was formulated in order to concentrate on the convective contributions of the carotid artery and internal jugular vein, the primary conduits for transfer of heat to and from the brain from an ice bath surrounding the neck. This "countercurrent flow heat exchanger" problem was formulated for a geometry based on an idealized neck anatomy derived from www.vesalius.com rather than from the discrete NLM/ Karlsruhe data. This was done in part to simplify the problem and in part to adjust the anatomy of the overdeveloped neck of the cadaver source of the NLM data. Thus, a separate three-dimensional mesh was generated for this idealized geometry of the neck in order to emphasize temperature gradients in the vicinity of the carotid artery and internal jugular vein, which are relatively close to the neck surface. The head and neck computational mesh, thermal and perfusion parameters and numerical results are available in animation, graphical and tabular form at the following web site

http://maido.uta.edu/~brian/research/head/head.html

Table 1. Physical Properties of Different Tissues Used in Thermal Modeling of a Human Head and Neck

Tissue Type	Conductivity (k) ($W\,m^{-1}\,{}^{o}C^{-1}$)	Specific Heat (C_p) ($J\,kg^{-1}\,{}^{o}C^{-1}$)	ρ Cp ($J\,m^{-3}\,{}^{o}C^{-1}$)*10^6	Perfusion (M_b) ($kg\,m^{-1}\,s^{-1}$)	Metabolic Rate (q_m) ($W\,m^{-3}$)
Skin	0.34	2495.0	3.7	0.433	33.0
Fat	0.21	2495.0	3.7	0.3	33.0
Muscle	0.52	3543.0	3.7	0.433	33.0
Bone	1.16	1500.0	2.4	0.066	5.0
Brain matter	0.52	3680.0	3.7	8.6	525.0
Eye	0.52	-	3.7	0	0
Liquor	0.52	-	3.7	0	0
Lens	0.34	-	3.7	0	0
Neural tissue	0.52	-	3.7	0	0
Optic nerve	0.52	-	3.7	0	0
Cartilage	0.52	-	3.7	0	0
Mucous membrane	0.0252	-	0.991	0	0
Glands	0.52	-	3.7	0	0
Air	0.0252	-	0.991	0	0
Blood	0.5		4.3	0	0

II. FINITE ELEMENT NUMERICAL FORMULATION

An approximate temperature distribution throughout a perfused tissue can be found by solving the bioheat transfer equation [1].

$$\rho C_p \frac{\partial \theta}{\partial t} = \nabla \cdot (k \nabla \theta) + M_b C_p (\theta_a - \theta) + q_m \qquad (2)$$

Here θ_a is the arterial perfusion temperature and q_m is the heat source due to metabolism. In this work we assume temperature-independent material properties and heat source terms. Under these conditions analytic solutions can be found only for very simple geometries. For realistic problems, we must resort to numerical approximations of the bioheat transfer equation. A finite element method, which is described here, was used [11]. In the finite element method, the domain where the solution is sought is divided into a finite number of parts called elements. In this work tetrahedral shaped elements are used exclusively. Almost any complex 3-D geometry, including a human head, can be decomposed into a finite number of well-shaped tetrahedral elements.

Applying the method of weighted residuals to (2) over a single tetrahedral element Ω^e with weight function v results in

$$\int_{\Omega^e} v\left\{\rho C_p \frac{\partial\theta}{\partial t} - k\left(\frac{\partial^2\theta}{\partial x^2}+\frac{\partial^2\theta}{\partial y^2}+\frac{\partial^2\theta}{\partial z^2}\right) - M_b C_p(\theta_a-\theta) - q_m\right\} d\Omega^e = 0 \quad (3)$$

Integrating (3) by parts once creates the weak statement for the element

$$\int_{\Omega^e} v\rho C_p \frac{\partial\theta}{\partial t} d\Omega^e + \int_{\Omega^e} k\left(\frac{\partial v}{\partial x}\frac{\partial\theta}{\partial x}+\frac{\partial v}{\partial y}\frac{\partial\theta}{\partial y}+\frac{\partial v}{\partial z}\frac{\partial\theta}{\partial z}\right)+vM_b C_p \theta d\Omega^e = \int_{\Omega^e} v(M_b C_p \theta_a + q_m) d\Omega^e - \int_{\Gamma^e} v(\vec{q}\cdot\vec{n}) d\Gamma^e \quad (4)$$

Here, Γ^e is the element of the surface. Variation of the temperature across the element can be expressed by

$$\theta(x,y,z) = \sum_{i=1}^{m} N_i(x,y,z)\theta_i \quad (5)$$

Here, i is an element local node number and m is the total number of element nodes, and N is the shape function associated with node i. Using Galerkin's method, the weight function v and the interpolation function for θ are chosen to be the same. Now define the matrix $[B]$ as

$$[B] = \begin{bmatrix} \frac{\partial N_1}{\partial x} & \frac{\partial N_2}{\partial x} & \cdots & \frac{\partial N_m}{\partial x} \\ \frac{\partial N_1}{\partial y} & \frac{\partial N_2}{\partial y} & \cdots & \frac{\partial N_m}{\partial y} \\ \frac{\partial N_1}{\partial z} & \frac{\partial N_2}{\partial z} & \cdots & \frac{\partial N_m}{\partial z} \end{bmatrix} \quad (6)$$

and the weak statement (4) can be written in the matrix form as

$$[C^e]\{\frac{\partial\theta^e}{\partial t}\} + [K_c^e]\{\theta^e\} = \{R^e\} \quad (7)$$

The local stiffness matrix, $[K_c^e]$, the local capacitance matrix, $[C^e]$, and right hand side vector, $\{R^e\}$, are evaluated as

$$[K_c^e] = \int_{\Omega^e} k[B]^T[B] + M_b C_p \{N\}\lfloor N \rfloor d\Omega_e \quad (8)$$

$$[C^e] = \int_{\Omega^e} \rho C_p \{N\}\lfloor N \rfloor d\Omega_e \quad (9)$$

$$\{R^e\} = \int_{\Omega^e} (q_m + M_b C_p \theta_a)\{N\} d\Omega_e - \int_{\Gamma^e} q_s \{N\} d\Gamma^e \quad (10)$$

for each element in the domain and then assembled into the global system of linear ordinary differential equations.

$$[C]\{\dot{\theta}\} + [K_c]\{\theta\} = \{R\} \quad (11)$$

This global system of equations can be solved to obtain the approximate solution to the bioheat transfer equation of the entire domain.

Here the time derivative of temperature, $\{\dot{\theta}\}$, is discretized using a finite difference approximation in time. The Crank-Nicolson algorithm is used and the final linear system of algebraic equations can be written as

$$\left(\frac{1}{2}[K_c]+\frac{1}{\Delta t}[C]\right)\{\theta_{n+1}\} = \left(-\frac{1}{2}[K_c]+\frac{1}{\Delta t}[C]\right)\{\theta_n\} + \frac{1}{2}(\{R\}_{n+1}+\{R\}_n) \quad (12)$$

where the subscript $n+1$ denotes the current time step and n denotes previous time step.

The approximate solution approaches the exact solution as the number of elements is increased and as the time step size Δt is reduced. However, increasing the number of elements and increasing the number of time steps also increases the computational cost. Therefore, the analyst must choose a mesh and time step size that will give acceptable results in a reasonable amount of computational time.

The finite element method was implemented in a C++ code. The code was written to be computationally efficient so that realistic simulations can be performed on a typical personal computer. The linear algebraic system is solved at each time step using an iterative sparse matrix solver based on conjugate gradient method with an incomplete factorization for preconditioning [12]. The solution for the previous time step is used as the initial guess for the iterative solver on the next time step, which provides enhanced converge rates. A typical time-accurate numerical simulation can be performed on a notebook PC (PIII 700 MHz) in less than 20 minutes using less than 64 MB of in-core memory [13,14,15].

III. NUMERICAL RESULTS FOR HEAD COOLING

Head Cooling With Ice Surface Temperature

In the FEM simulation of cooling of a realistic three-dimensional human head, only the face, eyes and nose were assumed to be at the steady room temperature of 25.0 °C. The base of the head was at a steady temperature of 37.0 °C. In all cases, the temperature was monitored at six points within the brain. The locations of the monitor points are shown in Figure 2. The simulation was terminated when all six points reached the target temperature of 33.0 °C.

The following four test cases were numerically analyzed:

Case 1. No tissue perfusion and θ_{surf} = 0.0 °C at t = 0 for the regions covered by a garment maintained at ice temperature.

Case 2. Tissue perfusion at internal elements, using reasonable values adapted from the UT Southwestern rhesus monkey experimental hypothermia series [5,6]. The same surface element boundary conditions as in Case 1 and $\theta_{art} = 35.0$ °C.

Case 3. A repeat of Case 2 with $\theta_{art} = 33.0$ °C.

Case 4. A repeat of Case 2 with $\theta_{art} = 31.0$ °C.

CASE 1. Surface at ice temperature, no perfusion cooling

As shown in Figure 3, without perfusion cooling of the brain, all tissues within the skull cool very slowly. Brain temperatures do not reach the 33.0 °C target within 30 minutes of the onset of surface cooling (Figure 4), although the brain did reach this target in 70 minutes (Figure 5). This result suggests that surface-only cooling at $\theta_{surf} = 0.0$ °C would not be a viable method for protection of the brain.

The total energy removed via the skin at 70 minutes is 386kJ, or a total average power of 92 watts (Figure 6). The peak power (as estimated by backward extrapolation close to t = 0) is over 200 watts. The absolute peak power value (at t = 0) was not calculated, since this modeling method requires extremely small mesh sizes and time increments near the surface and at t = 0, and unduly burdens the model. Furthermore, the instantaneous power level at the sudden onset of $\theta_{surf} = 0.0$ °C is unrealistic. In reality a $\theta_{surf} = 0.0$ °C boundary condition at the skin would be established over a short period of time (seconds), thus decreasing the peak power. Thus, it is not unreasonable to extrapolate the instantaneous power level backwards to the origin from values at slightly longer times.

CASE 2. Surface at ice temperature, perfusion cooling with $\theta_{art} = 35.0$ °C

In this case the brain still does not reach the 33.0 °C target within 30 minutes of combined cooling. In fact, a significant portion of the brain does not reach 34.0 °C within 30 minutes (Figure 7). The reason for this is that the $\theta_{art} = 35.0$ °C counteracts the conduction-based cooling from the surface. The total energy removed from the surface is 1700 kJ (Figure 8), which is irrelevant, since the 33 °C was not reached.

CASE 3. Surface at ice temperature, perfusion cooling with $\theta_{art} = 33.0$ °C

With cooled arterial blood at $\theta_{art} = 33.0$ °C the brain reaches the 33.0 °C target after approximately 7.5 minutes (Figures 9, 10 and 11). A total of 94.7 kJ is removed at an average rate of 210 watts and a near-initial peak power of 250 watts (Figure 12). This means that less total energy needs to be removed than in Case 1, but the average and peak powers are higher, due to the additional cooling load of incoming blood to the brain.

CASE 4. Surface at ice temperature, perfusion cooling with $\theta_{art} = 31.0$ °C

With blood pre-cooled at $\theta_{art} = 31.0$ °C the brain reaches the 33.0 °C target at approximately two minutes (Figure 13). A total of 44.8 kJ had to be removed at an average power of 344 watts and a near initial peak power of 250 watts (Figure 14).

It is important to remember in Cases 2, 3 and 4 that the energy load associated with cooling the arterial blood to the specified temperatures is not included. Assuming the arterial blood enters the head/neck region at the normal temperature of $\theta_{art} = 37.0$ °C, with an arterial blood flow rate of 250.0 ml/min (which is less than normal), the additional average power load for 60 minutes of neck cooling is 20 watts.

Cases 1-4 suggest that brain cooling to the target level of 33.0 °C within 30 minutes is feasible *if* the arterial supply can be accessed and the arterial temperature reduced to these levels (at least $\theta_{art} = 33.0$ °C). Thus, efforts should be made to cool the arterial blood, examining all possible modalities. By inference, these results suggest that the brain may cool to the target level in time, if the head surface temperature $\theta_{surf} = -15.0$ °C is applied.

The perfusion-assisted results of Cases 2-4 agree in general with the experimental results of the UT Southwestern study of primates [5,6] and underscore the importance of the blood-based convective mechanism in tissue cooling. The midline temperature profiles in the animations of Cases 3 and 4 indicate more rapid cooling of the cortical and subcortical regions than deeper structures. This is expected, since the conduction pathway is shorter to these regions than to the deeper structures, and there is some lateral conduction owing to the curvature of the skull.

A principal assumption of this analysis is that the perfusion distribution is uniform for all brain substructures, albeit significantly higher than for non-cerebral structures.

Of course, the patient with a cerebral ischemic event would not benefit from perfusion-cooling of the region of ischemia. This would imply that convective cooling would not be available, and that the region most in need of the protection of lower temperature would not receive it. However, brain perfusion distribution is quite volatile in healthy subjects, with blood rapidly shifted to regions experiencing high metabolism, as elegantly demonstrated by functional SPECT and MRI studies. This raises the important question of the regulation of blood supply to regions adjacent to the ischemic region: how, if at all, is the perfusion of these regions affected? Speculating for the moment that perfusion is unaltered in the adjacent regions; one may envision a potentially dangerous region of warm ischemia, surrounded by better-perfused, cooler tissues. In contrast to the results of cerebral hypoxia modeling, thermal diffusion is much higher than oxygen diffusion. Thus, a larger region might be afforded partial metabolic protection, *even if* oxygen cannot be delivered.

It should also be noted that perfusion cooling in the vicinity of an ischemic area would *enhance* ischemic region cooling, so long as the blood perfusate is at a lower temperature than the ischemic tissue. Conversely, if the blood perfusate is at a higher temperature than the conduction-cooled tissue, such as might occur in the vicinity of a near-surface ischemic region, the surface cooling effect would be blunted by the warm blood perfusate.

Head Cooling With a Convection Cooled Helmet

In this sequence of FEM simulations of brain cooling, a realistic three-dimensional human head was assumed covered by a cooling helmet [1]. The inner surface of the helmet was subjected to convection cooling boundary condition. The face, eyes and nose were uncovered and maintained at room temperature of 25.0 °C.

The following three test cases were analyzed:

Case 5. No tissue perfusion and surface heat transfer coefficient h_{surf} = 25.0 W m^{-2} °C^{-1} (derived from [1]).

Case 6. Tissue perfusion with θ_{art} = 37.0 °C at all internal elements of the head. The perfusion distribution was identical to this of Case 2.

Case 7. All other conditions were the same as in Case 6, but h_{surf} = 46.0 W m^{-2} °C^{-1} (the highest value observed in [1]).

CASE 5. Surface convection cooling via helmet applied at t = 0, with a circulating cooling fluid at 5.5°C. No tissue perfusion.

Temperature profiles at six points within the brain (Figure 2) are given in Figure 15. At 30 minutes the outer cortical positions (C, D, E) have cooled below 35.0 °C. However, the inner cortical regions are only beginning to cool. The heat extracted from the head (input power) calculated for that portion covered by the cooled garment is shown in Figure 16. The total heat extraction over 60 min is 150 kJ.

CASE 6. The surface convection conditions of Case 5, with θ_{art} = 37.0 °C tissue perfusion.

The temperature evolution profiles at the six sites in the brain are shown in Figure 17. There is negligible cooling (less than 0.1°C) at all sites at 30 minutes. Brain cooling is less efficient when a helmet is interposed between the scalp and the cold source. This is more evident when the cooling fluid is at 5.5°C rather than the ice temperature of Case 2, and even more evident when no pre-cooling of the cerebral arterial blood is assumed, also as in Case 2. These results emphasize the conclusions of Cases 1-4, that cooling of cerebral arterial blood (performed at sites other than the head) is necessary.

Case 7. The surface convection coefficient is increased to 46 W m^{-2} °C^{-1}, the highest value extracted from the cited experimental study [1]. Other conditions as in Case 6.

Almost doubling the convective heat transfer coefficient of Case 6 provides negligible cooling of the brain (Figure 18).

IV. NUMERICAL RESULTS FOR NECK COOLING

Due to the need to cool the blood perfusing the brain, it was decided to numerically simulate independent means of cooling of the neck. The objective was to learn if neck-only cooling would be sufficient to reduce the arterial blood temperature before it enters the brain. The neck-only geometric model was created where cylindrical neck, cervical vertebral, tracheal and carotid arterial and jugular venous dimensions were estimated from the www.vesalius.com website data. Due to symmetry, only half the neck was modeled (Figure 19) with different material regions shown in Figure 20.

In the neck model, the chest portion of the neck was set to 37.0 °C except at the vein outlet. The top of the neck model used an insulating thermal boundary condition except for the vein inlet and artery outlet.

Carotid arterial entry temperature was θ_{art} = 37.0 °C. Jugular vein entry temperature was arbitrarily set at 33.0 °C, to simulate target temperature cooling of the brain achieved from other sources (helmet, pharynx, axilla, etc.). For all cases a steady, fully developed blood flow was assumed in both the artery and vein. The peak arterial blood velocity was varied from high-normal (V_{max} = 0.5 m s^{-1}) to very low (V_{max} = 0.05 m s^{-1}) as might be assumed in a severe stroke.

The following two test cases were simulated numerically.

Case 8. No tissue perfusion was assumed and θ_{surf} = 0.0 °C was applied at the neck surface thus simulating countercurrent heat exchange between jugular vein and carotid artery.

Case 9. Case 6 for the unsteady state. θ_{surf} = 0.0 °C at t = 0 at the neck surface. No tissue perfusion.

CASE 8. Steady state ice cooling of the neck with countercurrent heat exchange between jugular vein and carotid artery.

Figure 21 depicts the steady state temperature distributions on a slice through the neck that goes through both artery and vein. The cooling front does not reach the artery in these views. It is blocked by the countercurrent heat exchange from the warmer 33°C down-coming venous blood. At the elevation of the base of the skull (14 cm) the centerline carotid arterial blood has still not begun to cool below 37.0 °C. However, the bulk average blood temperature shows a modest impact of neck cooling, despite the blocking effect of the countercurrent exchange. The average computed temperature in the artery for various locations in the neck is shown in the table below.

Table 2. Average Blood Temperature in the Neck Artery at Different Locations

Neck Elevation (cm)	Average Temperature (°C)
2.0	36.6
6.0	36.0
10.0	35.6
14.0	35.5

Table 3. Average Outlet Blood Temperature in the Neck Artery for Different Peak Blood Velocity

Peak Velocity (m/s)	Average Outlet Temperature (°C)
0.50	35.3
0.25	34.3
0.10	31.7
0.05	28.0

However, it is important to note that tissue perfusion at arterial temperature which would warm the blood, was not included in this simulation. Figure 22 depicts the steady state artery outlet temperature profile for various values of peak blood velocity in the artery and vein.

The simulations show that the blood in the artery can be cooled significantly if the blood velocity is reduced from normal by a factor of ten.

CASE 9. Case 8 for the transient cooling situation with θ_{surf} = 0.0 °C at t = 0 at the neck surface.

Numerical results show that power consumption for a neck-only cooling device is high, as was the case both for direct icing and convective cooling of the head. The peak power, estimated by backward extrapolation is approximately 200 W; while average power over 60 minutes is approximately 20 W.

It should be pointed out that the peak value for the carotid blood average speed is around V = 0.5 m/s. It is likely even higher than that considering the range of mass flow rates quoted in the literature. For the case of peak blood speed V = 0.5 m/s, the computed results show only slight cooling of the blood at the artery outlet. This agrees with theoretical and experimental data since a peak blood speed V = 0.5 m/s is most often used. The purpose of using different peak average blood speeds in the simulations was to show what could be achieved if the blood flow rate could be reduced somehow.

V. DISCUSSION

Our group has previously applied the Finite Element Method to simulate transient brain cooling in a subhuman primate model, the macaque rhesus monkey [5,6]. Cooling was driven either by ice packs applied to the scalp, or by direct exchange with arterial blood via the heat exchanger in a heart-lung machine, or by both means. While the mesh was less exactly prepared than in the current work, the predictions of that cooling simulation agreed well with direct experimental measurements at multiple sites within the brain. The FEM model was also calibrated, in the present work, with a problem for which a closed form solution is available: excellent agreement was obtained. Therefore we have full confidence in the accuracy of the current FEM results.

The perfusion term in the Pennes equation is $M_bC_p(\theta - \theta_{art})$ where M_b is the *capillary perfusion rate*, measured in ml. blood flow per minute per ml. of tissue. This "rate" can be understood as blood flow through the millions of capillaries in the tissue, proceeding in all directions at once, and thus not having a vector form. Furthermore, the extremely small capillary diameters and the relatively low blood velocity in the capillaries ensure that the blood temperature equilibrates with that of the immediately surrounding tissue before the blood leaves the capillaries. The magnitude of this perfusion heat flow term is, in general, much larger than the conduction heat flow term for temperature gradients achievable in human thermal physiology. Comparison of conduction-only *versus* perfusion plus conduction heat exchange in an experimental animal series amply demonstrates this important point [5,6].

Therefore, only by artificially cooling the arterial blood was it possible to cool the brain in the specified time (Cases 3, 4). Even pre-cooling arterial blood to θ_{art} = 35.0 °C did not achieve 33.0 °C global cooling of the brain within 30 minutes! The helmet with 5.5°C coolant did no better when the protective effect of warm perfusion was included; the core of the brain was not cooled within 30 minutes for any condition (Cases 6 and 7). Ice bath cooling of the neck, provided a small measure of carotid arterial precooling of the brain (θ_{art} = 35.5 °C average arterial temperature at 30 minutes). However, the model did not include tissue perfusion with warm blood, which would have substantially reduced the cooling effect.

VI. CONCLUSIONS

The simulations of cooling methods considered to date all indicate that no single means of external cooling of the head or neck is sufficient to reach the 33.0 °C target temperature in a reasonable period of time. Specifically, for the iced surface of the head, inclusion of the warm blood perfusion term effectively blocks the cooling wave from advancing beyond a few millimeters at 30 minutes.

The central conclusion of insubstantial cooling is supported by the modest cooling enhancements reported in other experimental investigations of externally applied cooling [1,16,17]. While experiments might be conducted with optimized cooling conditions, such as shaving the head, applying closer-fitting head and neck garments, convective enhancement of the head or neck ice baths, none of these means are likely to change the conclusion in conservative application of the method.

In simulations for head subjected to both constant surface temperature and convection, the face, eyes, and nose were assumed to be at the steady room temperature which might not be realistic considering the three-dimensional heat transfer. It should not be too difficult to incorporate a natural convection boundary condition on these surfaces in the model. However, since the results are dominated by the effect of the blood perfusion and not the surface boundary conditions in these small areas, changing the boundary conditions would not change the overall results significantly.

All of this suggests that other cooling means should be explored, including: 1) combinations of currently considered external cooling methods; 2) other external methods not yet addressed; 3) internal cooling methods [18].

There are other appealing external cooling methods. Ice conditions can be applied to the axillae, and groin, regions with good coupling to arterial and venous supplies. The turbinate sinuses and pharynx, proximal to the undersurface of the frontal brain, can be convectively cooled with oxygen.

Finally, there are invasive methods, including swallowing of cold fluids, and blood access technologies which might be considered. These are less desirable because of technical considerations and the advanced degree of knowledge and training that would be required. As shown by a similar calculation, the cooling effect of 4.0 °C injected saline (1 to 2 liters) would be much larger and more immediate [18] than that of any test case studied in this paper. However, this cooling effect would be transient, not continuous, and it would require an invasive procedure.

NOMENCLATURE

C_p	Coefficient of specific heat
h_{surf}	Thermal convection coefficient
k	Thermal conduction coefficient
M_b	Capillary perfusion rate
q	Surface heat flux
q_m	Metabolic heat source
t	Time
V	Average blood speed
x,y,z	Cartesian coordinates
ρ	Density of the living tissue
θ	Temperature
θ_{art}	Artery blood temperature
θ_{surf}	Surface temperature

ACKNOWLEDGMENTS

The authors are grateful to Dr.-Ing. Frank B. Sachse, Institute of Biomedical Engineering, University of Karlsruhe for providing a digitized geometry of a real human head.

REFERENCES

1. Ku, et.al., (1996) "Analytical Study of the Cooled Helmet Human Data," *American J. Phys. Med Rehab.* vol. 75, pp. 443.
2. Proceedings of First Users Conference of the National Library of Medicine's *Visible Human Project*, October 7-8, 1996, National Institute of Health, Bethesda, MD.
3. Sachse, F., Werner, C., Müller, M., Meyer-Waarden, K., (1996) "Preprocessing of the Visible Man Dataset for the Generation of Macroscopic Anatomical Models", Proc. First Users Conference of the National Library of Medicine's Visible Human Project.
4. Sachse, F., Werner, C., Müller, M., Meyer-Waarden, K., (1996) "Segmentation and Tissue-Classification of the Visible Man Dataset Using the Computertomographic Scans and the Thin-Section Photos", Proc. First Users Conference of the National Library of Medicine's Visible Human Project.
5. Olsen, R. W., (1985) "Temperature Profiles in the Head and Other Tissues of the Macaque Rhesus Monkey Subjected to Surface and/or Core Cooling," Doctoral Dissertation, University of Texas Southwestern Medical Center at Dallas, Dallas, Texas.
6. Olsen, R. W., Hayes, L. J., Wisler, E. H., Nikaidoh, H., and Eberhart, R. C., (1985) "Influence of Hypothermia and Circulatory Arrest on Cerebral Temperature Distributions," *ASME J. of Biomechanical Eng.*, vol. 107, pp. 354-360.
7. *Heat Transfer in Medicine and Biology: Analysis and Applications*, Appendix II, vol. II (1985) Plenum Press.
8. *Heat Transfer in Medicine and Biology*, vol. 1, chap. 12, pp. 279-283.
9. Bowman, H. F., Cravalho, E. G. and Woods, M., (1975) "Theory, Measurement, and Application of Thermal Properties of Biomaterials," *Annual Review of Biophysics and Bioeng.*, Vol. 4, pp. 43-80.
10. Valvano, J. W., Cochran, J. R., and Diller, K. R., (1985) "Thermal Conductivity and Diffusivity of Biomaterials Measured with Self-Heated Thermistors," *International Journal of Thermophysics*, Vol. 6, No. 3, pp. 301-310.
11. Huebner, K. H., Thorton, E. A., and Byrom, T. G., (1995) *The Finite Element Method for Engineers.* John Wiley and Sons, New York, NY, third edition.
12. Balay, S., Gropp, W. D., McInnes, L. S., and Smith, B. F., (1999) "PETSc 2.0 Users Manual," Technical Report ANL-95/11 – Revision 2.0.24, Argonne National Laboratory.
13. Dennis, B. H. and Dulikravich, G. S., (1999) "Simultaneous Determination of Temperatures, Heat Fluxes, Deformations, and Tractions on Inaccessible Boundaries," *ASME Journal of Heat Transfer*, Vol. 121, pp. 537-545.
14. Dennis, B. H., and Dulikravich, G. S., (2000) "Determination of Unsteady Container Temperatures During Freezing of Three-dimensional Organs With Constrained Thermal Stresses," in *Internat. Symposium on Inverse Problems in Engineering Mechanics – ISIP'2k*, (eds: M. Tanaka and G. S. Dulikravich), Nagano, Japan, March 7-10, 2000, Elsevier
15. Dennis, B. H., Dulikravich, G. S., and Rabin, Y., (2000) "Optimization of Organ Freezing Protocols With Specified Allowable Thermal Stress Levels," ASME IMECE 2000, Orlando, FL, Nov. 5-10, 2000, HTD-Vol. 368/BED-Vol. 47, pp. 33-48.
16. Corbett, R. J. T. and Laptook, A., (1988) "Failure of Localized Head Cooling to Reduce Brain Temperature in Adult Humans," *Neuro Report*, vol. 9, pp. 2721-2725.
17. *Amer. J. Phys. Med. Rehab. vol.* 75, pp. 443, 1996.
18. Bernard, S. A., Jones, B. M., and Horne, M. K., (1997) "A Clinical Trial of Induced Hypothermia in Comatose Survivors of Prehospital Cardiac Arrest", *Ann. Emerg. Med.*, vol. 30, pp. 146-153.

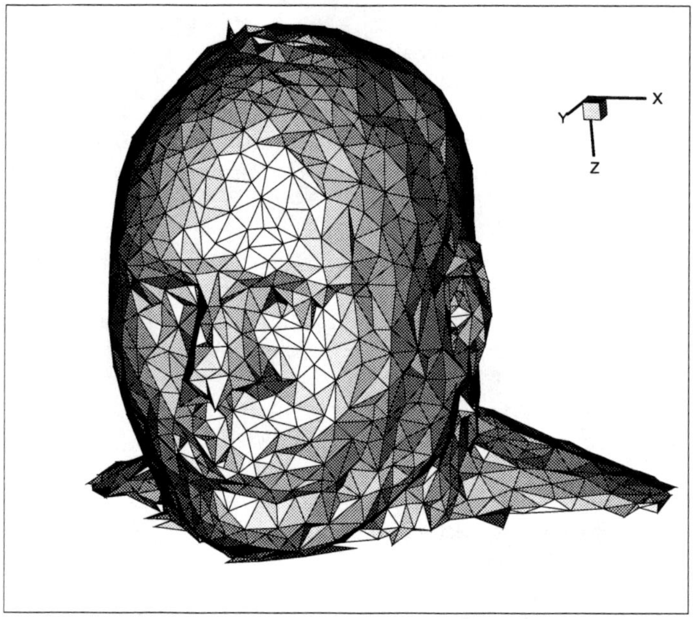

Figure 1. View of the triangular surface mesh of the realistic human head [3,4].

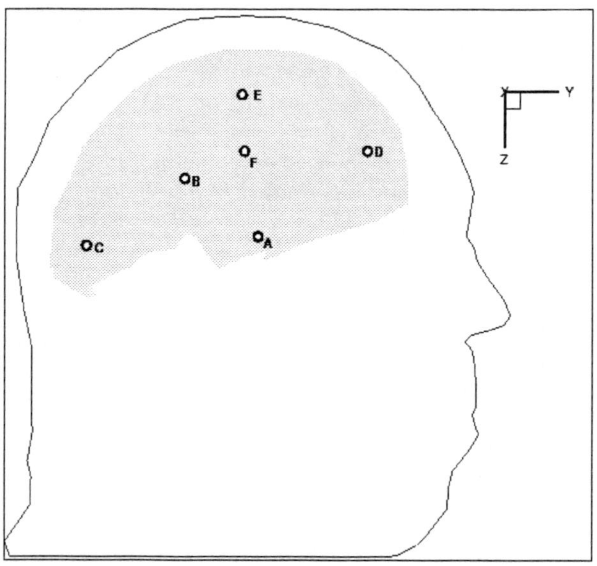

Figure 2. Temperature monitor points on a vertical symmetry slice taken in the middle of head.

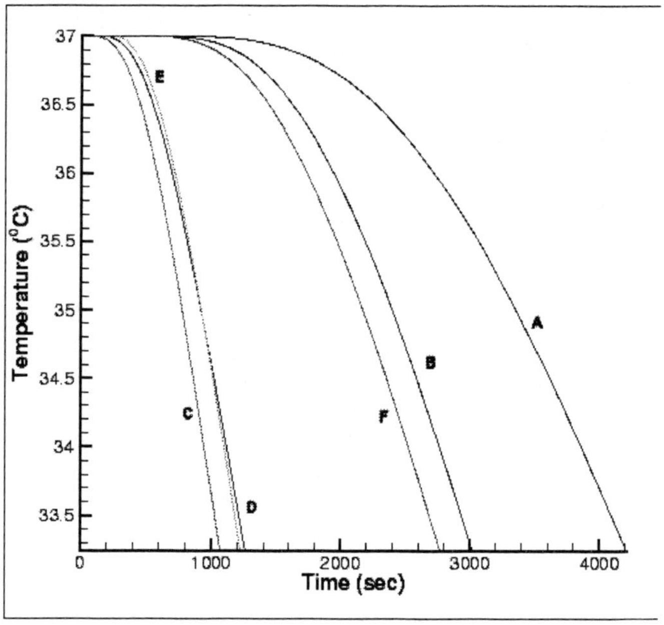

Figure 3. Temperature evolution at monitor points in the brain for Case 1.

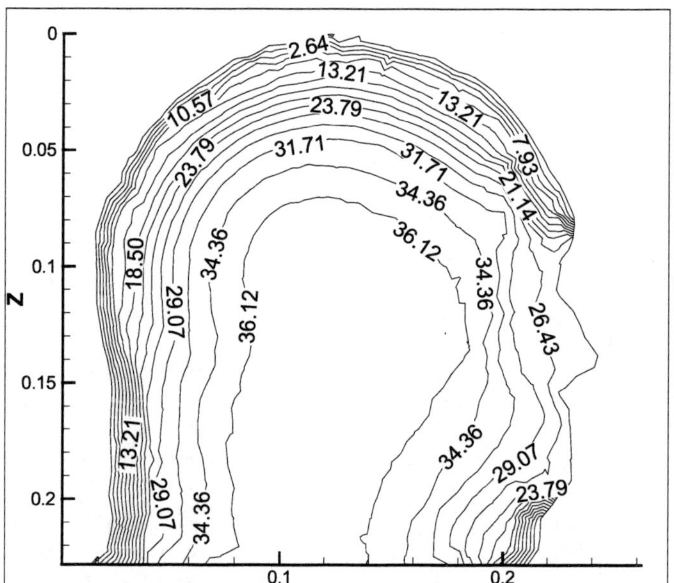

Figure 4. Temperature contours on slice through middle of head for Case 1 after 30 minutes of cooling.

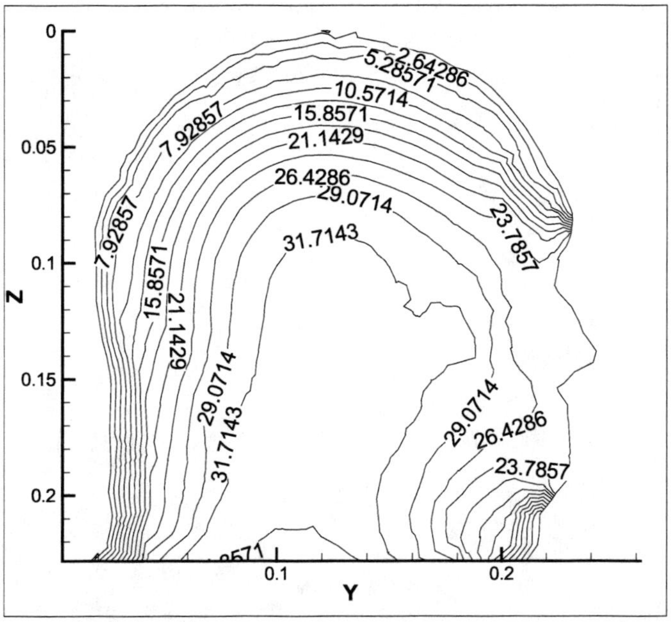

Figure 5. Temperature contours on slice through middle of head for Case 1 after 70 minutes of cooling.

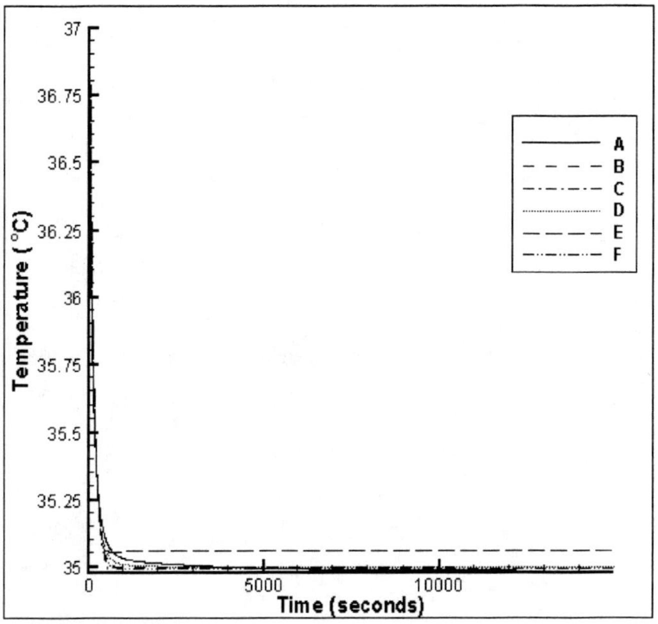

Figure 7. Temperature evolution at monitor points in the brain for Case 2.

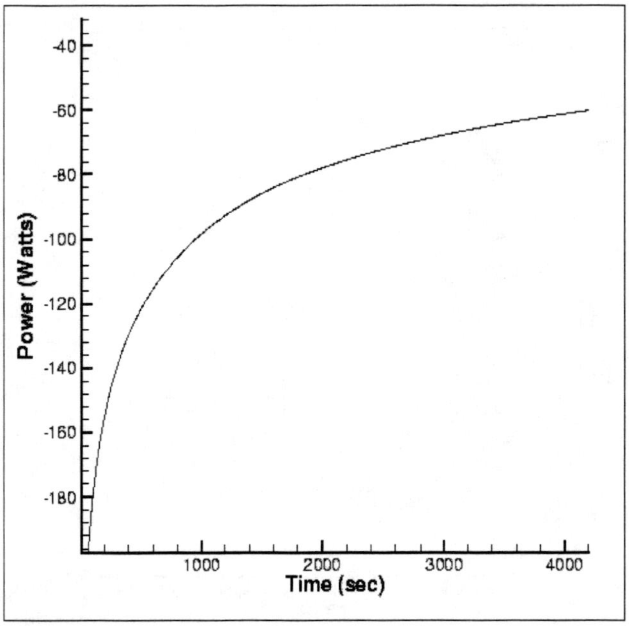

Figure 6. Power transferred across 0.0 °C surface of the head for Case 1.

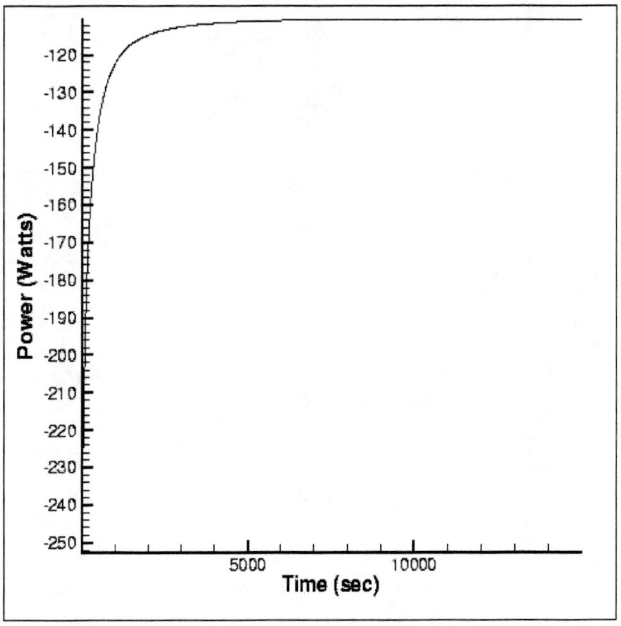

Figure 8. Power transferred across 0.0 °C surface of the head for Case 2.

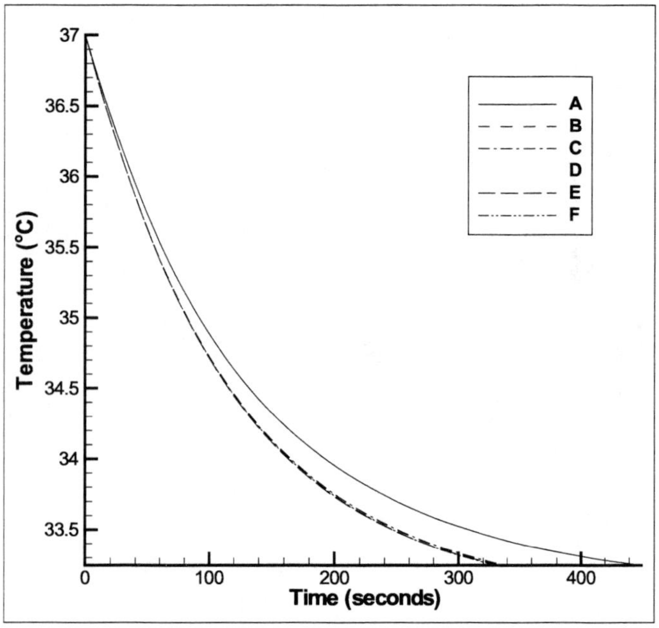

Figure 9. Temperature evolution at monitor points in the brain for Case 3.

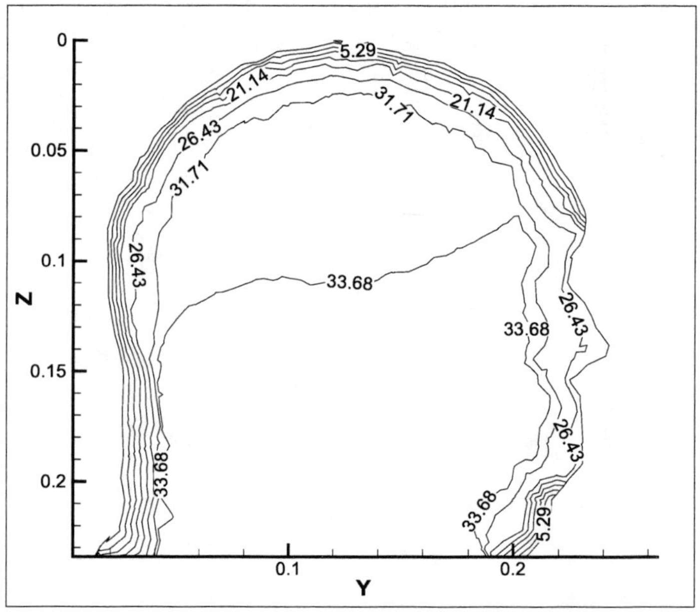

Figure 11. Temperature contours on slice through middle of head for Case 3 after 8 minutes of cooling.

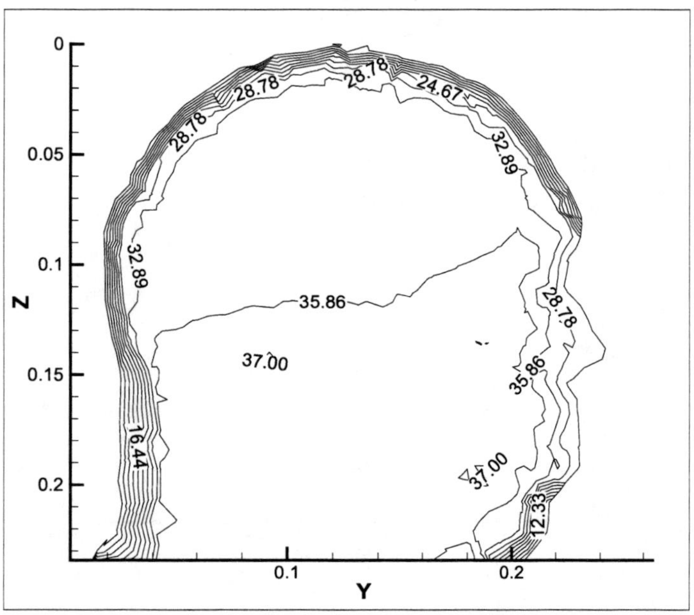

Figure 10. Temperature contours on slice through middle of head for Case 3 after 4 minutes of cooling.

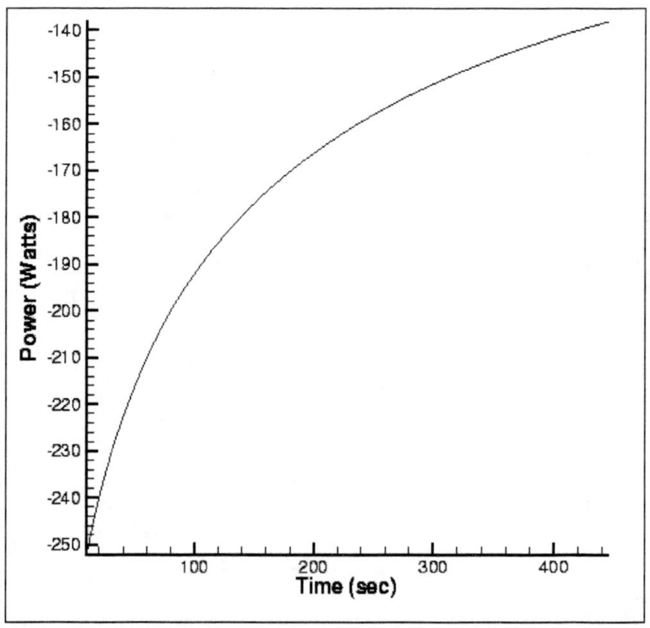

Figure 12. Power transferred across 0.0 °C surface of the head for Case 3.

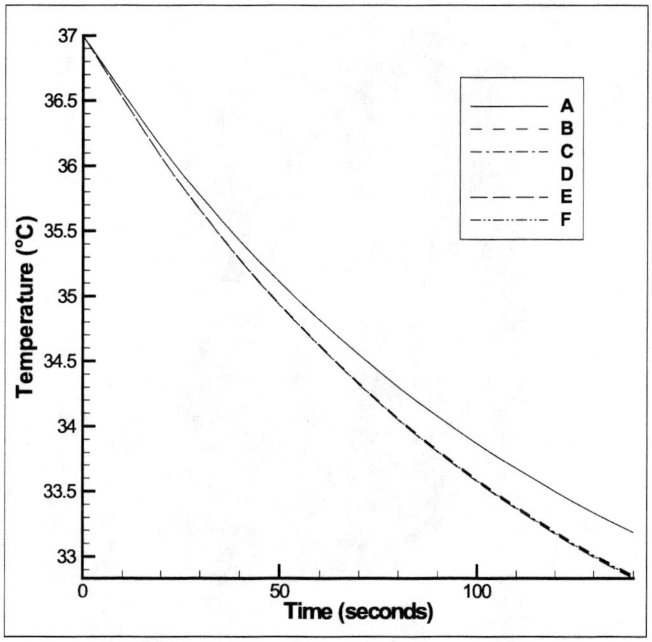

Figure 13. Temperature evolution at monitor points in the brain for Case 4.

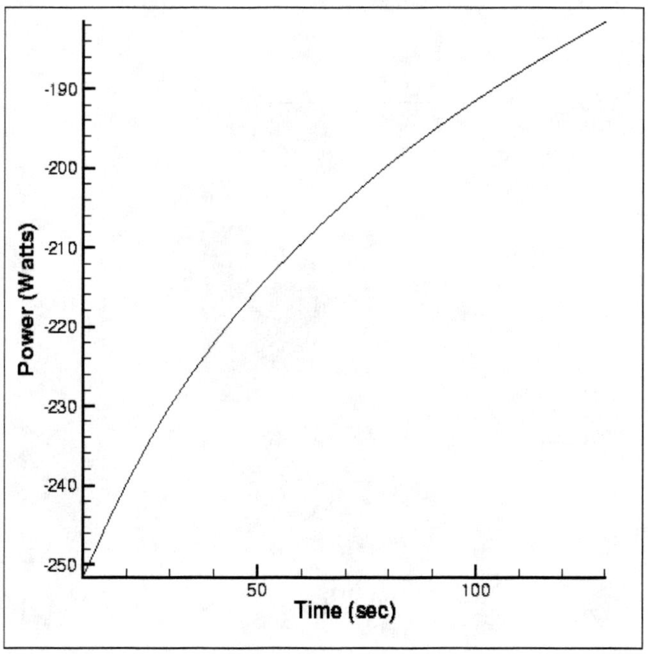

Figure 14. Power transferred across 0 °C surface of the head for Case 4.

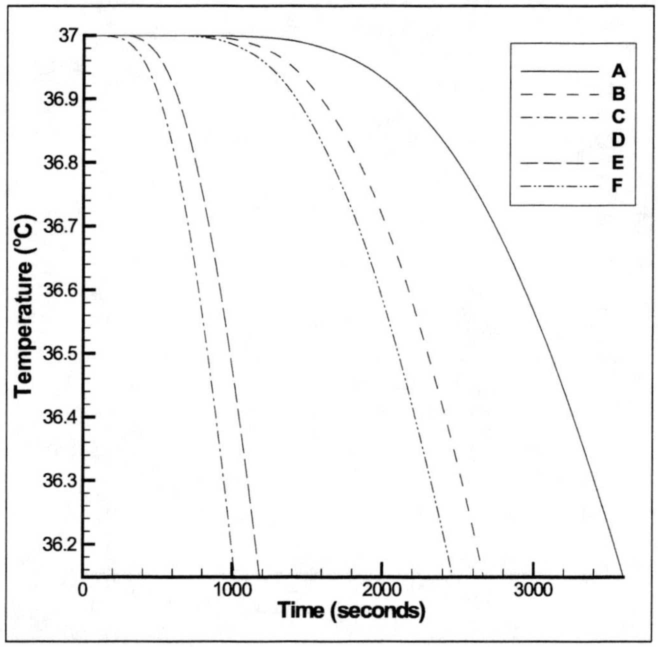

Figure 15. Temperature evolution at monitor points in the brain for Case 5.

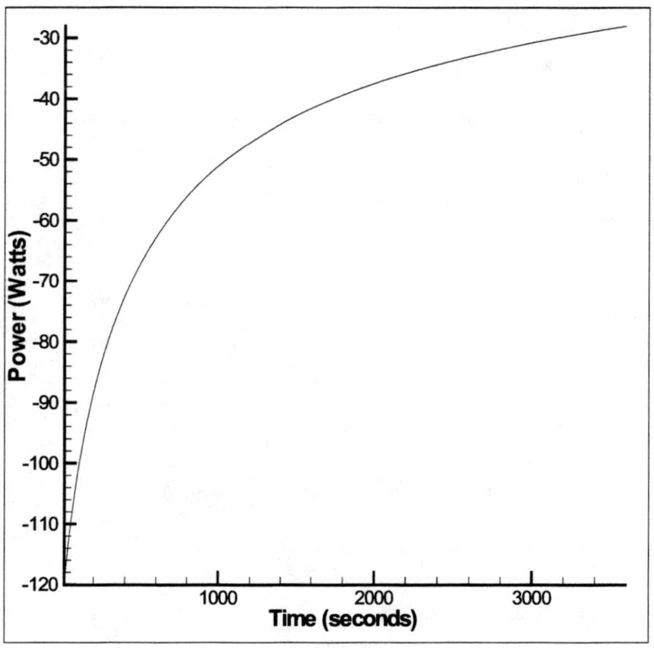

Figure 16. Power transferred across 0°C surface of the head for Case 5.

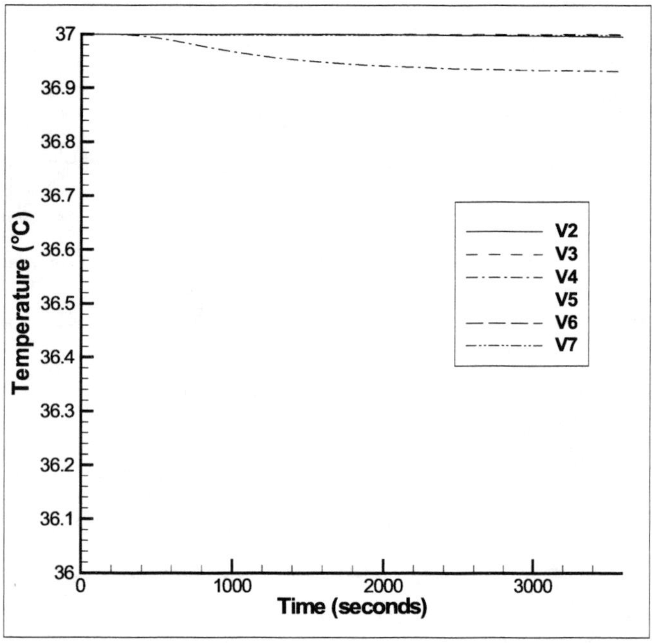

Figure 17. Temperature evolution at monitor points in the brain for Case 6.

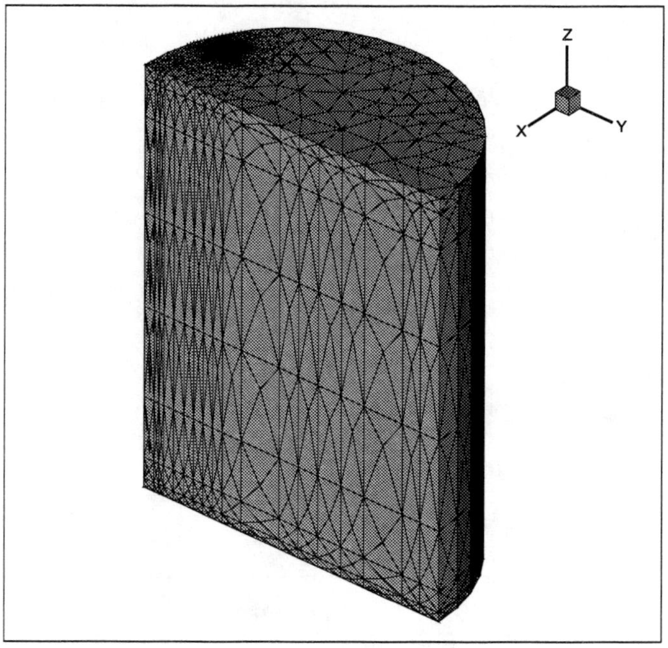

Figure 19. Triangular surface mesh for neck model.

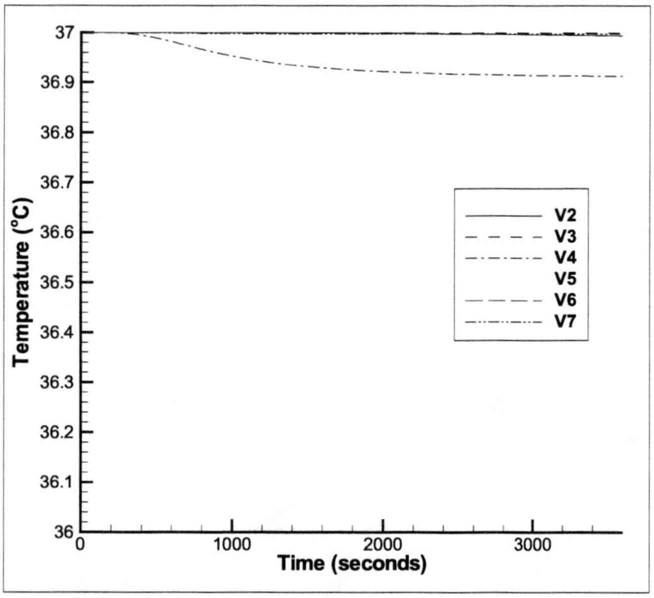

Figure 18. Temperature evolution at monitor points in the brain for Case 7.

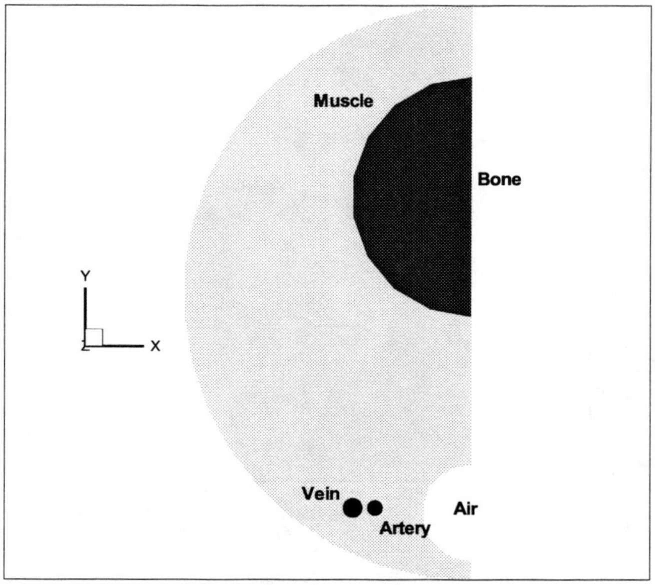

Figure 20. Material regions and geometry for the idealized neck model.

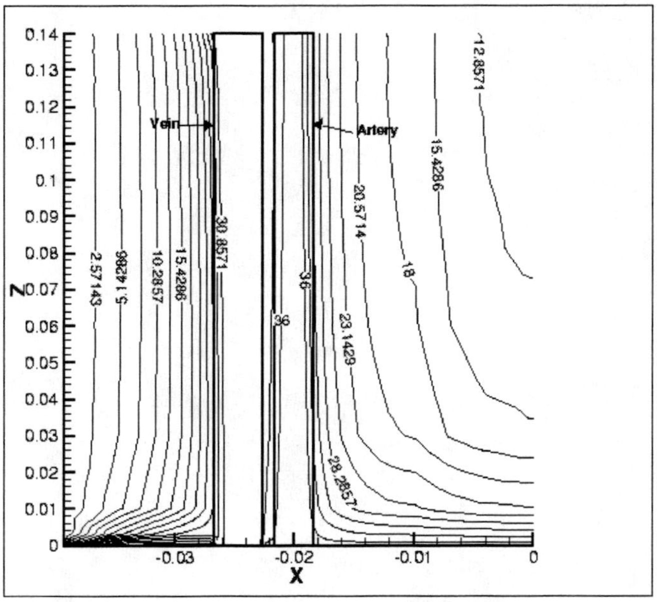

Figure 21. Steady state isotherms on neck slice that passes through both vein and artery (V_{max} = 0.5 m/s).

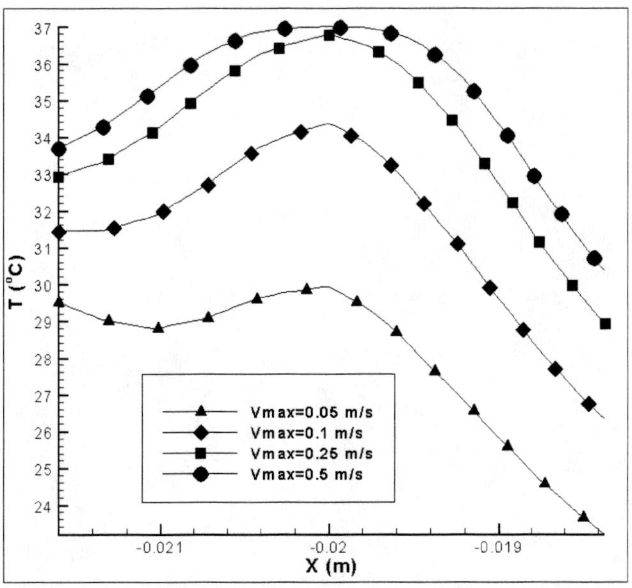

Figure 22. Artery outlet temperature profile for various values of peak blood velocity in artery and vein.

Proceedings of IMECE2002
ASME International Mechanical Engineering Congress & Exposition
November 17–22, 2002, New Orleans, Louisiana

IMECE2002-32046

FINITE ELEMENT ANALYSIS OF RADIO-FREQUENCY ABLATION IN A RECONSTRUCTED REALISTIC HEPATIC GEOMETRY

Ajit K. Pandey
Department of Mechanical Engineering
University of Cincinnati
Cincinnati, OH 45221-0072
pandeyat@email.uc.edu

Isaac Chang
Center for Devices and Radiological Health
U. S. Food and Drug Administration
Rockville, Maryland 20852
iac@cdrh.fda.gov

Matthew R. Myers
Center for Devices and Radiological Health
U. S. Food and Drug Administration
Rockville, Maryland 20852
mrm@cdrh.fda.gov

Rupak K. Banerjee
Department of Mechanical Engineering
University of Cincinnati
Cincinnati, OH 45221-0072
rbanerje@gauss.mie.uc.edu

ABSTRACT

Radio-frequency (RF) ablation is a minimally invasive procedure that has the potential for widespread use in hepatic cancer therapy. In the procedure, RF current is applied to the tissue, resulting in the conversion of electrical to heat energy and thus, a rise in temperature, with the goal of eventual tumor necrosis. Potential complications from the procedure include insufficient heating of large tumors, resulting in tumor recursion, as well as excessive thermal damage to healthy tissue. Mathematical models are valuable in predicting the temperature rise within the organ during RF ablation, thereby enhancing the success rate of the procedure. Eventually, models can be used to guide ablation procedures, by predicting the optimal set of operational parameters e.g., catheter probe geometry and placement, given patient-specific information. The present study focuses on the analysis of temperature rise within a reconstructed model of a realistic three-dimensional (3D) section of a porcine liver during RF ablation.

This study calculates the effect of blood flow through arteries as well as perfusion through the liver on the time-dependent temperature distribution near the RF ablation probe (Figure 1). For a time duration of 30 min of an ablation procedure, a temperature of about 80°C could be achieved over a diameter of about 4 cm with the present RF probe. As an initial step, the present study includes isotropic hepatic tissue and blood properties.

INTRODUCTION

The term "radiofrequency ablation (RF) probe" refers to a class of medical devices operating between 460-550 kHz that delivers therapeutic energy into soft tissues.

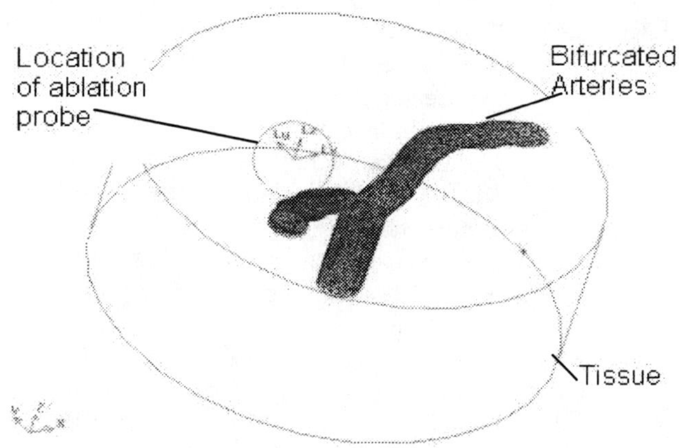

Figure 1: Reconstructed bifurcated arteries with surrounding hepatic tissues. The RF ablation probe is inserted in the spherical tissue domain, considered to be the ablation zone.

Typically, energy is injected through the probe, into soft tissues, to a ground electrode that is usually placed on the surface of a patient, thereby completing the electric circuit. Although radiofrequency ablation has been widely used for several decades, significant advancement in understanding the physics has only recently been made.

In computing heat transfer during ablation procedures, the effect of blood perfusion is indirectly modeled as a source term using a lumped heat-transfer coefficient in the energy equation. This coefficient is assumed to apply throughout the entire liver [1, 2]. While such an approach is sufficient for heat transfer through an ensemble of capillaries, it is unlikely to be accurate in the case of ablation near a large blood vessel. In this study, both convective blood flow and heat transfer within the conjugate domains of large vessels and surrounding porous hepatic tissues are solved directly using the coupled mass, momentum, and energy transport equations, thus, improving on the previous studies. This direct-simulation approach requires information regarding the vascular geometry of the treated organ. MRI scans of the affected tissue are acquired, followed by segmentation and generation of a computational mesh. The technique will be illustrated below for the case of an excised porcine liver model.

Accurate analysis of RF ablation procedures also requires a realistic representation of the electric field generated by the RF catheter. In principle, this field is coupled to the temperature and fluid-flow fields, as the electric conductivity is temperature dependent. For the present study, as an initial step, electric field values which have been previously determined are equated to a distributed heat source either as a constant or, more accurately, as a gaussian function over a tissue domain. The temperature rise over time induced by the heat source near a large blood vessel within the porcine liver is calculated, and the effect of blood perfusion on the temperature field is assessed.

NOMENCLATURE

c_p	Specific heat, J/kg.K
k_i	Permeability, m^2
k_e	Thermal conductivity, W/m.K
p	Fluid pore pressure, N/m^2
q_s	Heat source from ablation probe, W/m^3
T	Temperature, ^0C
u	Velocity, m/s
u	Average velocity, m/s
V	Volume, m^3
μ	Dynamic viscosity, Ns/m^2
?	Density, kg/m^3
φ	Porosity

subscripts:

e	Effective property of both the fluid and solid
f	Fluid properties
i, j, k	x, y, & z direction for a 3 – D problem
s	Solid matrix
t	Total

METHODOLOGY

The analysis of the RF ablation procedure is accomplished through a Finite-Element (FE) solution to the fluid-flow and energy equations within the reconstructed liver geometry having blood flow through the arteries and blood perfusion through the porous hepatic tissue. This section describes the methodologies for acquiring the physiologic geometry and for numerically representing the transport processes.

Image Reconstruction and Meshing: Using a small bore MRI, a total of 128 images spaced 625 microns apart are obtained from an excised porcine liver segment. In order to identify the geometry of the arteries and surrounding tissue, images are segmented by automatic upper and lower threshold values [3], which are based on image contrast. All gray colored pixels between the threshold values are treated as hepatic tissue, thus identifying the artery supplying blood to the liver. 3D reconstruction of the geometry is achieved by a region growing technique from the segmented MR scans. After checking the smoothness of the 3D geometry, polylines with different colors are created to distinguish the segmented arteries and tissue surfaces. Subsequently, surface and curve fit to the polylines are incorporated and an IGES format of the geometry containing real vertices, edges, and surfaces are imported for automatic, linear, tetrahedral mesh generation [4]. The excised liver section is of a cylindrical shape (Figure 1) with a height of 72 mm and diameter of 180 mm, containing about 190,000 elements.

Equations: In the present study the hepatic tissue is modeled as a porous medium with blood perfusion. Within porous materials, there are two distinct regions of interest: fluid and solid. If V_f represents the volume of fluid and V_s represent the solid volume, such that the total volume, V_t is given by,

$$V_t = V_f + V_s. \quad (1)$$

Porosity is defined by,

$$\phi = V_f / V_t \quad (2)$$

In order to construct a continuum model, the volume elements are considered sufficiently large compared with the length scale of an individual pore. With these assumptions and steady state conditions, the conservation of mass is expressed by the continuity relation,

$$\nabla \cdot \bar{u} = 0 \quad (3)$$

where ∇□is the divergence operator and $\bar{u}$ is the average of the fluid velocity over V_t. The momentum equation is expressed by Darcy's law,

$$\bar{u} = -k \nabla p \quad (4)$$

where k is the hydraulic conductivity tensor, $k_i = k\mu$ is the permeability, and ∇ is the gradient operator. k is dependent on the pore geometry and interstitial matrix of the porous medium and is a measure of the conductance of the material to fluid flow. An implicit assumption is that the ambient temperature does not change the fluid density, and subsequently the fluid flow profile.

The following energy equation for porous media is also used for temperature distribution analysis:

$$(\rho c_p)_e \frac{\partial T}{\partial t} + \rho c_p u_j \frac{\partial T}{\partial x_j} = \frac{\partial}{\partial x_j}\left(k_e \frac{\partial T}{\partial x_j} \right) + q_s \quad (5)$$

The effective properties are related to fluid and solid matrix properties by the relations:

$$(\rho c_p)_e = \phi \rho c_p + (1-\phi)(\rho c_p)_s \quad (6)$$

$$k_e = \phi k + (1-\phi) k_s$$

(7)

Where properties without subscripts are those of the fluid. First, the momentum equation is solved to calculate the velocity field in the liver tissue. Subsequently, the velocity field is used as input and coupled with the energy equation to get the temperature field in the liver domain at different time. Within the arterial region, the mass, momentum and energy equations of fluid are applied. The governing equations are solved [5] with a segregated solver and a second order trapezoidal time integration scheme.

Model Parameters: In the present study the RF probe is modeled as a spherical heat source with an average value of 2.7×10^5 W/m^3 over a volume of 14.1×10^{-6} m^3. The boundary condition imposed on all external surfaces is zero heat flux except at the inlet of the artery and the tissue near it, where a constant temperature of 37°C is used. The average blood perfusion rate through the hepatic tissue is maintained at 0.6 ml/min/ml volume of liver [1,2], which, in turn, resulted in a blood velocity of 0.124 m/s is for the reconstructed inlet diameter of the artery. A stress free boundary condition is used at all other boundaries.

Tissue and Blood Properties: The present study considered isotropic tissue and blood properties at 37°C. The tissue and blood properties [6] are: density, ? = 1050 kg/m^3; capacity, c_p=3600 J/kg.K; conductivity, k = 0.502 W/m.K. The infinite shear-rate blood viscosity, μ=0.0034 Ns/m^2. As an initial estimate the hepatic tissue porosity is considered to be 0.2 and the isotropic permeability is 10^{-10} m^2.

RESULTS AND DISCUSSION

Figure 2 represents the temperature variation in x-direction with time. As expected, there is a rapid increase in temperature during initial times, e.g., 8°C increase in first 2 min whereas only 5°C increase occurs between 30 min and 60 min. For a time duration of 30 min. of ablation procedure, a temperature of about 80°C could be achieved over a diameter of about 4 cm with the present RF probe.

The effects of blood flow rate variation, different probe design, and direct coupling of electrical field with transport equations are presently under evaluation. Anisotropic hepatic tissue properties including permeability and temperature dependent thermal conductivities will also be evaluated. The impact of more detailed reconstructed geometry including hepatic arteries, portal veins, hepatic veins, hepatic sinuses surrounded the tissue will be included in future studies.

We would like to thank Jan Johannessan of the Center for Vetenary Medicine, US Food and Drug Administration, for his assistance in attaining the MRI images used in this study.

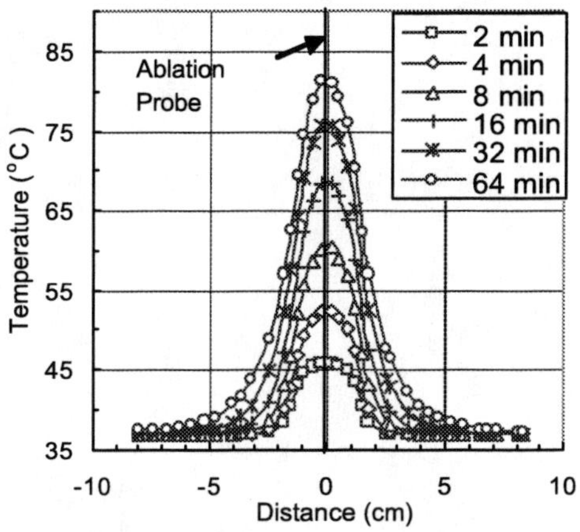

Figure 2: Temporal variation of temperature distribution in x-direction

REFERENCES

1. Tungjitkusolmun S., Staelin S.T., Haemmerich D., Tsai J-Z, Cao H., Webster J.G., Lee F.T., Vorperian V., and Mahvi M., "Three dimensional finite-element analyses for radio-frequency hepatic tumor ablation", IEEE, Vol. 49, pp. 3-9, 2002.
2. Tungjitkusolmun S., Woo E.J., Tsai J-Z, Cao H., Webster J.G., and Vorperian V., "Thermal-electrical finite element modeling for radio frequency cardiac ablation: effects of changes in myocardial properties", Med. and Bio. Engg. & Computing, Vol. 8, pp. 562-568, 2000.
3. Mimics User Manual, ver. 7.1, 2001, Materialise Inc., Ann Arbor, MI.
4. Gambit User Manual, ver. 1.3, 2002, Fluent Inc., Lebanon, NH.
5. Fidap User Manual, ver. 8.6, 2002, Fluent Inc., Lebanon, NH.
6. Duck F., "Physical properties of tissue", Academic Press, New York, 1990, pp. 167-223.

Proceedings of IMECE2002
ASME International Mechanical Engineering Congress & Exposition
November 17-22, 2002, New Orleans, Louisiana

IMECE2002-32047

POTENTIAL FOR THERMAL INJURY DURING SURGICAL OSTEOTOMIES

James W. Allen[1] Darren L. Hitt[1] Marc R. Sarnow[2]

[1]Department of Mechanical Engineering
University of Vermont
Burlington, Vermont

[2]Department of Surgery
University of Vermont
Burlington, Vermont

ABSTRACT
Medical literature indicates that permanent tissue damage can occur when exposed to temperature elevations of as little as 5°C for times as short as a few seconds. Clinically, it has also been observed that during surgical osteotomies (bone cutting procedures) significant amounts of heat are typically generated. We present experimental data that suggests critical conditions for thermal injury are regularly exceeded during simulated surgical procedures using cadeveric metatarsal bones.

INTRODUCTION
Inherent in the mechanical cutting of any material is heat generation through friction. In the case of bone, the excess heat can lead to a condition where the periosteum, which supplies blood to the bone, is damaged and can no longer sustain a healthy and normal regeneration of the injured bone. Avascular thermal necrosis is a potentially serious condition in which fat deposits at the site of the thermal injury causing abnormal fusing and/or no fusing of the cut site. Osteotomies are very common surgical procedures performed on thousands of patients daily throughout the world. As such, determining the mechanical conditions under which excessive heat is generated can help to lessen or even eliminate the occurrence of thermal injury.

The serious potential for thermal injury can be quantified as follows, bearing in mind the normal human body temperature of 37°C. The medical literature reports that exposure to temperatures as low as 42°C can cause vascular injury, but will not deter bone repair. Bone regeneration may be significantly impaired when the periosteum is exposed to temperatures of 47°C for 1 minute and bone tissue is permanently damaged at 50°C and 1 minute exposure.[1-3]

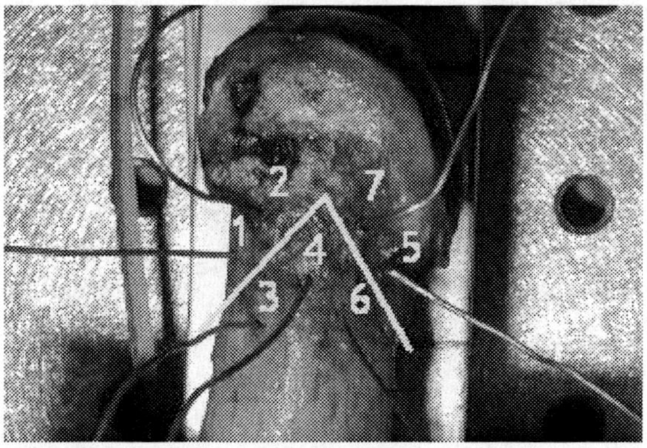

Figure 1- A cadaveric first metatarsal head instrumented with thermocouples set at a depth of 0.5 mm. The lines indicated the path of the surgical cut. The average thermocouple distance from the cut lines 2.5mm +/- 1mm

EXPERIMENT
An experimental apparatus was developed to investigate levels of heat generation during simulated surgery. Cadaveric bones were used in order to preserve human bone properties and morphology; these were supplied by the Podiatry Institute (Tucker, Georgia). First metatarsals were chosen and cut using a standard Austin cut which is used in common bunion surgery. An array of seven K Type thermocouples were press fit in the bone to a depth of 0.5mm (Figure 1). Cuts were made using a Hall® saggital saw (Fletcher Allen Health Care Burlington, VT). The bones were mounted in a fiber-filled resin and held to a beam instrumented with an array of strain gages used to monitor the applied force during the cut. Thermocouple and strain gage data were recorded using a PC-based data acquisition system at a sampling rate of ~1.2 Hz per channel. Cortical bone thickness was measured as well as distance from thermocouple to cut at each site.

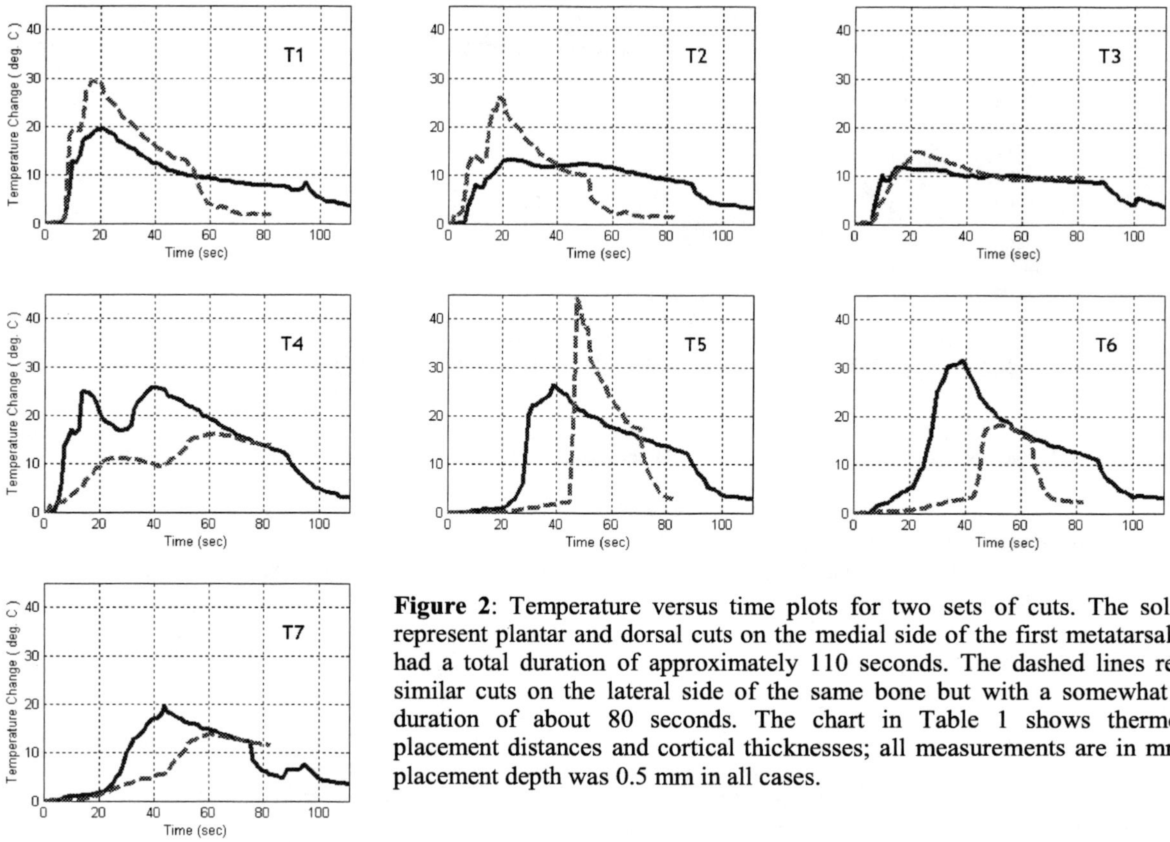

Figure 2: Temperature versus time plots for two sets of cuts. The solid lines represent plantar and dorsal cuts on the medial side of the first metatarsal, which had a total duration of approximately 110 seconds. The dashed lines represent similar cuts on the lateral side of the same bone but with a somewhat shorter duration of about 80 seconds. The chart in Table 1 shows thermocouple placement distances and cortical thicknesses; all measurements are in mm. The placement depth was 0.5 mm in all cases.

RESULTS

Data acquired from fourteen simulated surgeries suggests that critical temperatures may be regularly exceeded during typical procedures. A sample data set of temperature histories is shown in Figure 2. For this discrete data set, it is likely that both vasculature and bone tissue will be damaged at the measured locations. Clearly, bone tissue closer to the actual cut will experience even greater temperatures. Current efforts are directed toward obtaining a numerical model for the transient temperature field, which will permit a more complete estimate for the extent of region of thermal injury.

In summary, the results of the present work appear to support our hypothesis that thermally injurious conditions may be common in osteotomies. Having been so established, ongoing experimental work is being directed at identifying specific effects of key mechanical and physiological parameters such as force applied to bone, blade configuration, bone density, cortical thickness, irrigation and abnormal tissue.

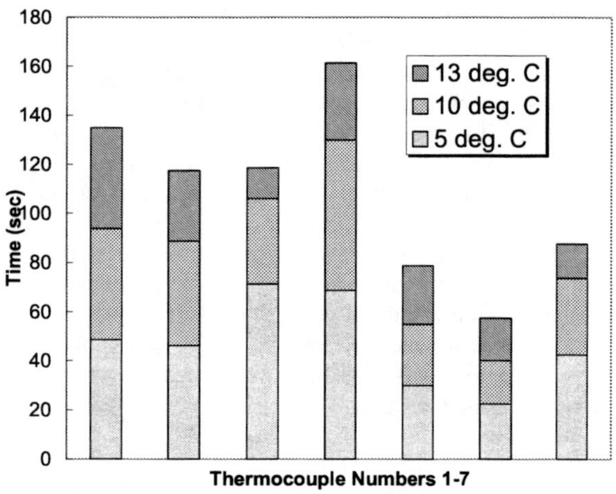

Figure 3 – Duration of critical temperature excesses based on Figure 2.

REFERENCES

1. R. Eriksson, *"Heat-Induced Bone Tissue Injury: An In Vivo Investigation of Heat Tolerance of Bone Tissue and Temperature Rise in the Drilling of Cortical Bone* [dissertation], Goteburg, Sweden, University of Goteburg, 1984.
2. A.R. Eriksson and T. Albrektsson, "Temperature threshold levels for heat induced bone tissue injury: a vital microscopic study in the rabbit," J Prosthet Dent, Vol. 50, p 101, 1983.
3. A. Eriksson, T. Albrektsson, B. Grane, et al., "Thermal injury to bone," Int J Oral Surg, Vol. 11, p 115, 1982.

	Cut 1+2 (Solid Line)			Cut 3+4 (Dasned Line)	
	Cortical Thickness	Distance From Cut		Cortical Thickness	Distance From Cut
T1	1.47	1.48	T1	1	1.49
T2	1.13	2.00	T2	0.89	2.59
T3	1.47	2.00	T3	1	1.53
T4	1.13	2.23	T4	1.08	1.63
T5	1.13	2.07	T5	1.18	0.65
T6	1.13	1.28	T6	1.18	1.48
T7	1.13	1.36	T7	1.08	2.12

Table 1

Proceedings of IMECE2002
ASME International Mechanical Engineering Congress & Exposition
November 17–22, 2002, New Orleans, Louisiana

IMECE2002-32048

INVESTIGATION OF THE THERMAL AND INJURY BEHAVIOR DURING MICROWAVE THERMAL THERAPY OF PORCINE KIDNEY

Xiaoming He[1], Shawn Mcgee[2,4], James E. Coad[2], Paul A. Iaizzo[3,5], David J. Swanlund[1], Stan Kluge[6], Eric Rudie[6] and John C. Bischof[1,3,4]

Department of Mechanical Engineering[1], Laboratory Medicine & Pathology[2], Biomedical Engineering[3], Urological Surgery[4] and Experimental Surgery[5], University of Minnesota and Urologix Inc. [6]

ABSTRACT

In this paper, we report on the characterization of microwave therapy of normal porcine kidneys both *in vitro* and *in vivo*. This technology is being developed for eventual use in the treatment of small renal cell carcinoma (RCC) by minimally invasive procedures. During experiments, microwave energy was applied through an interstitial microwave probe (Urologix, Plymouth, MN) to the kidney cortex with occasional involvement of the kidney medulla. The thermal histories at several locations were recorded. After treatment, the kidneys were bisected and small tissue slices were cut out at approximately the same depth as the thermal probes. The tissue slices were further processed for histological study. Both cellular injury and the area of microvascular stasis were quantitatively evaluated by histology. Absolute rate kinetic models of cellular injury and vascular stasis were developed and fit to this data. A 3-D finite element thermal model based on the Pennes Bioheat equation was developed and solved using a commercial software package (ANSYS, V5.7). The Specific Absorption Rate (SAR) of the microwave probe was measured experimentally in tissue equivalent gel-like solution. The thermal model was first validated by the measured *in vitro* thermal histories. It was then used to determine the blood perfusion term *in vivo*.

INTRODUCTION

Renal cell carcinoma (RCC) is the third most common genitourinary malignancy (after prostate and bladder cancer), accounting for an estimated 29,900 new cases and 11,000 deaths annually in the United States [1, 2]. Traditional treatments of RCC include watchful waiting, radical nephrectomy, partial nephrectomy, radiation therapy, and chemotherapy [3]. Watchful waiting is only applicable to patients with an expected life expectancy less than 10 years. Radical or partial nephrectomy is usually accompanied by severe morbidity and is not appropriate for senior patients. Both radiation therapy and chemotherapy have severe side effects [3]. Because of these disadvantages and the increased incidental discovery of smaller size and earlier stage RCC with modern imaging tools, many minimally invasive procedures have been proposed as alternatives to the traditional therapies. One such procedure is interstitial microwave thermal therapy. Although continued research is needed to determine the ultimate merits of this technique, interstitial microwave thermal therapy is a promising alternative treatment of RCC [2-5].

During interstitial microwave thermal therapy, microwave is irradiated from an antenna probe inserted into the diseased tissue. The interaction between the microwave and surrounding tissue triggers a rapid temperature increase in the tissue. Sufficient accumulation of the elevated temperature over time, (i.e., the supraphysiological thermal history) will reach cytotoxic levels, which leads to microvascular stasis, cell death, and finally tissue necrosis.

Several designs of the antenna probe exist including simple line dipole [6-8], multisection [9], and helical [10, 11]. The helical antenna has been shown to give the most homogeneous longitudinal specific absorption rate (SAR) pattern compared to the other designs [10, 12]. Another advantage of helical antenna is that their heating patterns are not significantly affected by insertion depth [11, 13]. When needed, several antennae can be used in an array with phase modulation to treat a large diseased tissue volume with complex anatomical geometry [14]. The antenna can be naked or embedded in a Foley-type catheter with water-cooling channels running through it [10, 15-16].

To predict the treatment efficacy of microwave thermal therapy, the heat generation, namely the SAR, must be determined first. Two approaches have been widely used to determine this SAR distribution. One is by measuring the initial rate of temperature rise in a tissue equivalent gel or gel-like solution following a short pulse of microwave energy from the probe [10, 17-19]. The SAR distribution can then be found by fitting the measured SAR at a limited number of points into a pre-assumed expression. Another way is to solve the Maxwell equations describing the electromagnetics during microwave thermal therapy [6, 8]. Although more universal, this method is very difficult to apply to probes with complex design like the helical antenna.

Several papers have been published trying to understand the thermal phenomena during microwave thermal therapy [8, 11, 17-18, 20-22]. In these studies, the Pennes bioheat equation [23] was numerically solved with measured SAR distribution to compute the temperature field. Other bioheat transfer models were not used, probably due to the difficulty in implementing those models [24]. Constant blood perfusion rate and thermal properties were frequently used [8, 17], although many studies have shown that the blood perfusion is temperature and time dependent [25-29] and the thermal properties (particularly the thermal conductivity) are temperature dependent [30]. Some studies did use a temperature and time dependent blood perfusion term in their model [21-22, 31]. Either the finite element method or the finite difference method was used to discretize the simplified tissue geometry of interest. Although many studies have shown that the cell/tissue injury behavior during thermal therapy is dependent on both the cell/tissue type and the thermal history experienced [30-33], a critical temperature of 54°C or 55°C was used as the criterion of complete tissue necrosis by some studies [11, 20], probably due to the lack of relevant injury parameters of their cell/tissue system. The thermal dose model [34] without accounting for differences in cell/tissue types or the Arrhenius injury model with the injury parameters from single cell studies or *in vitro* tissue studies was also used to predict the heat induced injury [11, 21-22,31, 35].

In this study, we focused our attention on the following two unanswered questions involved with microwave thermal therapy of normal kidney tissue: ① what are the kinetics of acute cellular injury and microvascular stasis? and ② what is the correlation between the dynamics of blood flow and degree of microvascular stasis?

MATERIAL AND METHOD

2.1 Microwave delivery apparatus

The microwave delivery apparatus is a product developed by Urologix Inc. (Plymouth, MN). The microwave generator can deliver up to 60W power to a microwave probe through a coaxial waveguide. The probe consists of a helical antenna and a Foley type plastic catheter. There are four separate water flow channels distributed circumferentially inside the catheter wall. The coolant water is maintained at constant temperature of 37°C by a refrigeration system built into the apparatus. The microwave probe has a 25.4mm active length and 4.74mm diameter (Fig. 1).

2.2 Probe positioning and thermal measurement

A specially designed jig (Fig. 1) was used to position the microwave probe and the fiber optic Luxtron thermometers, which will be referred to here as "thermal probes", in fixed reproducible positions in both the *in vitro* and *in vivo* kidneys. The jig has a large central hole through which the microwave probe was placed. Only the working part of the microwave probe was introduced from the middle top point into the kidney (Fig. 1). The thermal probes were placed through smaller holes each 5mm away from each other starting at the central hole. The insertion depth of the thermal probes was 12.0mm or about half of the length of microwave probe. Stainless steel needles with the same diameter as the microwave probe or the thermal probes were used to create paths through the kidney for both the microwave probe and the thermal probes, respectively, thus ensuring correct positioning. For *in vivo* study, seven thermal probes were used. One was placed on the surface of the microwave probe. The other six were placed at 5mm, 10mm and 15mm on each side of the microwave probe along the sagittal direction. For *in vitro* study, two additional probes were placed at 20mm. The above placements of the microwave and thermal probes are shown in Fig. 1.

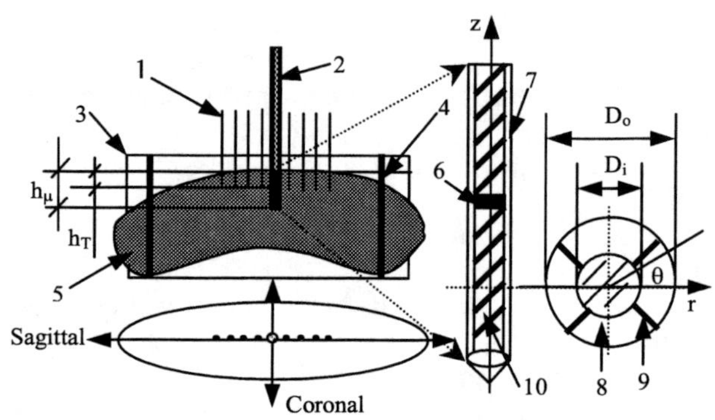

Fig.1 Schematic of probe positioning and the microwave probe. 1, Fiber optic thermal probes connected to Luxtron data acquisition system; 2, Microwave probe connected to microwave generator with built-in refrigeration system; 3, a special designed jig used to position the thermal probes accurately; 4, surgical bandage used to hold the kidney in the jig; 5, the kidney was either removed from the animal *in vitro* or left *in vivo*; 6, dipole junction; 7, catheter wall; 8, coolant flow channel; 9, extrusion (0.13mm thick); 10, antenna wire; h_μ, active probe length (25.4mm); h_T, thermal probe insertion depth; D_i, antenna coil diameter (2.54mm); D_o, probe diameter (4.74mm); z, axial direction; r, radial direction; θ, angular direction.

2.3 Animals preparation

Mongrel pigs weighing between 60kg and 90kg were cared for by the University of Minnesota Research Animal Resources Department in accordance with the principles set forth in the NIH guide for the Care and Use of Laboratory Animals, revised in 1996. The experimental protocol was reviewed and approved by the University of Minnesota Institutional Animal Care and Use Committee.

On the day of experiments, pigs were sedated with an intramuscular injection of 5-10 mg/kg of Telazol and 5-10 mg/kg Xylazine. An ear vein was cannulated for delivery of 5-10 mg/kg thiopental to aid in intubation. The animals were then intubated and placed on mechanical ventilation consisting of 60% air/40%oxygen. Maintenance anesthesia consisted of 1.2 MAC isoflurane. Once the animals were fully anesthetized, cutdowns were performed and catheters were placed in the right external jugular vein and the right carotid artery. The venous line was used for any further drug delivery and the arterial line was used for continuous blood pressure measurement.

2.4 In vivo thermal therapy

Kidneys were approached through a midline laparotomy incision and dissected free of connective tissue while allowing all vessels to continue to flow with blood. Kidneys were then fixed in the jig and the microwave and thermal probes were placed as shown in Fig. 1. The coolant water, which had the same temperature as the body baseline temperature (37 °C), was first allowed to circulate through the flow channels in the catheter of the microwave probe. The probe was used to deliver 50W of microwave power at 921MHz for 10 minutes. Kidneys were then allowed to cool down to baseline temperature, which lasted about 5 minutes. Subsequently, the microwave and thermal probes and the jig were removed. The abdomen of the animal was temporarily closed and the kidneys continued to be perfused for two hours. The animals were then euthanized and the kidneys were removed and processed for further injury measurement.

2.5 In vitro thermal therapy

The procedures for *in vitro* thermal therapy were similar to the *in vivo* case, except that 1) kidneys were removed from animals prior to therapy and 2) it took about 15min for the temperature to return to baseline temperature after therapy.

2.6 Injury measurement

All kidneys were weighed, measured and subsequently bisected. Each of the kidneys was then quartered and serially sectioned at 3 mm intervals to a depth of 21 mm from the insertion point of the microwave probe. Each tissue section contained a quarter of the thermal lesion's cross-section, including the entire lesion radius and a portion of the adjacent non-treated cortex. The fifth tissue layer (1.2-1.5 cm depth) was immediately fixed in formalin to observe the acute effects of thermal injury (microvascular level). The fourth tissue layer (0.9-1.2 cm depth) was placed into histology processing cassettes and tissue cultured at room temperature for 36-48 hours in RPMI Medium 1640 that was enriched with 1% Penicillin-Streptomycin and 2% fetal bovine serum. The use of 36-48 hours culture allowed observation of the comprehensive effects of thermal injury at the cellular level. The fourth section was then formalin-fixed for study of microvascular stasis. Both the fourth and fifth tissue layers were paraffin-embedded, sectioned, and slides stained with hematoxylin and eosin (H&E) for histological analysis. Cellular injury and microvascular stasis were then evaluated along with appropriate controls from unaffected adjacent renal parenchyma using stained sections. A total of ten kidneys (five for *in vivo* and five for *in vitro*) was studied. For each kidney, four tissue slices were analyzed and the average value of the four slices was used.

To quantify the lesion size of thermal injury, two different zones were defined at the cellular level, namely, the complete ablative zone (CAZ) and the partial ablative zone (PAZ). The cellular viability on the edge of the complete ablative zone was set at 20%, while 60% was set at the edge of partial ablative zone. Another two zones, the red zone (RZ) and the pink zone (PZ), were also defined at the microvascular level. The RZ was characterized by vascular congestion and red blood cell damage. The pink zone was characterized by proteinacious renal tubules. The edge of the red zone was defined as tissue in which approximately 20% of the microvasculature showed no stasis. Similarly, the edge of the pink zone was defined as tissue in which 60% of the renal tubules were free of the gelled proteinacious fluid. A schematic of the different zones is given in Fig. 2.

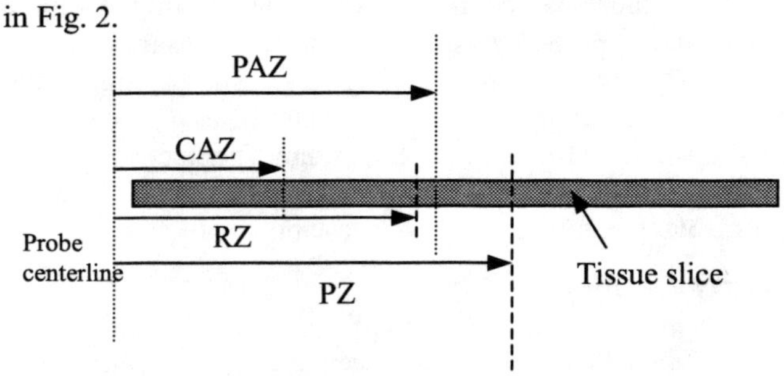

Fig. 2 Schematic of the different zones at the cellular and vascular level *in vivo*. Only PAZ and CAZ exist *in vitro*.

2.7 SAR measurement

The SAR measurement setup consisted of a special designed micromanipulator, a data acquisition system, the microwave delivery system, and a cylindrical container filled with microwave tissue equivalent gel-like solution whose major components include water and ethylene glycol. This solution is a modification from reference [36]. The microwave probe was fixed by two supports at the bottom and top of the cylinder. A tiny Luxtron fluoroptic thermal probe was assembled on the micromanipulator to allow free movement of the thermal probe around the microwave probe. For each measurement, the microwave power was turned on for ~2000ms and the thermal histories at the point of the thermal probe were recorded by the data acquisition system. The linear section of this thermal history, approximately between 700ms and 1000ms, was used to calculate the SAR at the corresponding point using the following equation:

$$SAR = C_p \, {\Delta T}/{\Delta t} \tag{1}$$

where Cp is the specific heat of the gel-like solution, ΔT is the temperature increase during the time period Δt. Although theoretically, the SAR pattern should be symmetric with respect to the dipole junction (Fig. 1), this symmetry was not used to reduce the experimental points in this study as the error of the probe placement might compromise the symmetry and hence the experimental results. Therefore, in this study, a total of 25×13×4 points (25 points with 12 points on each side of and one point on the dipole junction in the axial direction, 13 points in the radial direction and four points in the angular directions, viz, 0°, 90°, 180°, and 270°, refer to Fig. 1) were measured. The distance between adjacent points in the radial direction was 2mm. In the axial direction, the distance between points inside the probe length was 2mm, while this value was 4mm outside the probe length. The experimental results obtained were then averaged according to symmetry to overcome placement error. The flexible tolerance method [38-40] was used to perform the non-linear fitting of the following expression to the experimental results:

$$SAR = A\frac{e^{-2\alpha r}}{r^n}\left(1 + a|z| + bz^2 + c|z|^3 + dz^4\right)e^{(z/z_0)^2} \quad (2)$$

where r is the radius calculated from center of the microwave probe, z is the axial coordinate calculated from the dipole junction of the probe, α is the attenuation constant equal to 0.413 (cm^{-1}) [18] and A, n, a, b, c, d, and z_0 are fitting constants. This expression was selected based on several previous studies [10, 17-18]. By minimizing the root mean square between the experimental and the predicted results at all experimental points, the unknown constants can be found.

2.8 The cell injury model

2.8.1 The absolute rate theory model

The injury kinetics of the kidney cells during microwave thermal therapy were modeled as a first order irreversible rate process as follows [23]:

$$N \overset{K_c}{\Rightarrow} I \quad (3)$$

where N represents the natural state, I represents the injured state, and K_c is the rate of the transition from the N to I state. The fractional cell survival after treatment (FS) can be derived and given as follows:

$$FS = \exp\left(-\int_0^t K_c dt\right) \quad (4)$$

where t is time period during which the cells were exposed to supraphysiological temperatures. According to absolute rate theory, the rate of transition, K_c, can be modeled as follows:

$$K_c = \left(\frac{k}{h}\right)Te^{-\Delta G_c / RT} \quad (5)$$

$$\Delta G_c = \Delta H_c - T\Delta S_c \quad (6)$$

where k is the Boltzman constant (1.38×10^{-23}J/K), h is the Planck's constant (6.625×10^{-34} J/s), T is temperature in K, ΔG is the change of free energy (J/mole), ΔH is the activation enthalpy (J/mole), ΔS is the activation entropy (J/(mole·K)), and R is the universal gas constant (8.314J/(mole·K)). The subscript 'c' represents cellular level.

2.8.2 Procedure to find the ΔH and ΔS

The values of ΔH and ΔS were found from the results of thermal and injury measurements. The procedures involved two major steps. The first step was to find the critical thermal histories at the edge of each injury zone for all kidneys. This was done by Lagrangian interpolation using the measured thermal histories in several locations, which bracketed the location of interest. The second step involved a nonlinear fitting of Eqns. 4-6 using these thermal histories and corresponding injuries on the edge of each zone. This was done by minimizing the root mean square between the measured and the fitted injuries. The fitted injury was calculated using the interpolated thermal histories and Eqns. 4-6. The flexible tolerance method [38-40] was used to do the minimization. After these two steps, the optimal ΔH and ΔS were found for both the in *vivo* and *in vitro* cases.

2.9 The blood perfusion model

The blood perfusion, ω, was modeled as follows:

$$\omega = \omega_{bl}F(T,t) \quad (7)$$

where ω_{bl} is the baseline blood perfusion (measured to be 1.2±0.5ml/(g·min) by placing a flowmeter on the renal artery), F(T,t) is a temperature (T) and time (t) dependent function which is correlated to microvascular stasis. If we simplify the kinetics of microvascular stasis as a first order irreversible rate process, the fraction of tissue showing microvascular stasis, or the degree of stasis, DS, can be as follows:

$$DS = 1 - \exp\left(-\int_0^t K_s dt\right) \quad (8)$$

where K_s is the rate of microvascular stasis due to heating. K_s was calculated similarly as K_c as shown in Eqn. 5-6, but with ΔG_s, ΔH_s and ΔS_s. These ΔH_s and ΔS_s were found using the same procedures described in section 2.8.2, but the results from microvascular stasis analysis instead of cellular injury were used. Since blood perfusion is affected by many factors like physiological response, hydrostatic pressure, rheologic properties of blood [37], the function F(T,t) may not be as simple as Eqn. 8. Further, assuming that all the possible factors that may affect blood perfusion are correlated to microvascular stasis, blood perfusion can be only dependent on the degree of stasis. Now, Eqn. 7 can be rewritten as follows:

$$\omega = \omega_{bl}F(DS) \quad (9)$$

The function F(DS) will be discussed more in detail below.

2.10 The finite element bioheat transfer model

The Pennes bioheat equation [26] was used in this study to model the thermal distribution in the kidney,

$$\rho_t c_t \frac{\partial T}{\partial t} = k_t \nabla \cdot \nabla T + \omega \rho_b c_b (T_a - T) + \rho_t \cdot SAR \quad (10)$$

where ρ is density in kg/m^3, c is specific heat capacity in J/(kg·°C), k is thermal conductivity in W/(m·°C), T is temperature in °C, t is time in s, the symbol 'ω' represents blood perfusion rate in ml/(g·s), and SAR is the specific heat absorption in W/kg. The subscript 't' represents tissue, the subscript 'b' represents blood and the subscript 'a' represents artery. The thermal properties used are given in Table 1[30].

The boundary conditions used were as follows:
at the kidney surface:

$$- k_t \frac{\partial T}{\partial n} = h_1 (T - T_r) \quad (11)$$

at the probe surface:

$$- k_t \frac{\partial T}{\partial n} = h_2 (T - T_{cl}) \quad (12)$$

where n is the outward normal of the corresponding surface, T_r is the environmental temperature (25°C for *in vitro* and 33°C for *in vivo*), T_{cl} is the coolant temperature (37°C) , h_1 is the average heat transfer coefficient between the kidney surface and the environment, and h_2 is the average heat transfer coefficient between the kidney tissue and the coolant inside the flow channels of the catheter. By matching the measured *in vitro* thermal histories to the predicted thermal histories, h_1 was found to be 30W/(m^2·°C) and h_2 was found to be 450W/(m^2·°C) while the coolant was flowing and 10 W/(m^2·°C) when the coolant was not flowing. The boundary conditions on all the other surfaces were adiabatic.

Due to symmetry, only a quarter of the kidney was considered as the computational domain. A volume rendering method using spline fitting was adopted to create the 3-D solid model of the kidney using ANSYS (V5.7). This solid model was parameterized and varied with the kidney dimensions including the length (L) between the two poles, the height (H) and the thickness (T) (refer to Fig. 3). The dimensions used in this study were the average values from the direct measurements of the sizes of the kidneys used. Specifically, L=130±3mm, H=55±3mm and T= 35±2mm. The solid model was further discretized using ANSYS, as shown in Fig. 3.

Table 1 Thermal properties used in the FEM heat transfer model

Tissue density (ρ_t)	950 kg/m^3
Blood density (ρ_b)	1000 kg/m^3
Tissue thermal conductivity (K_t)	0.4976+0.001174*T W/(m·°C)
Tissue specific heat (c_t)	4000. J/(kg·°C)
Blood specific heat (c_b)	4100. J/(kg·°C)

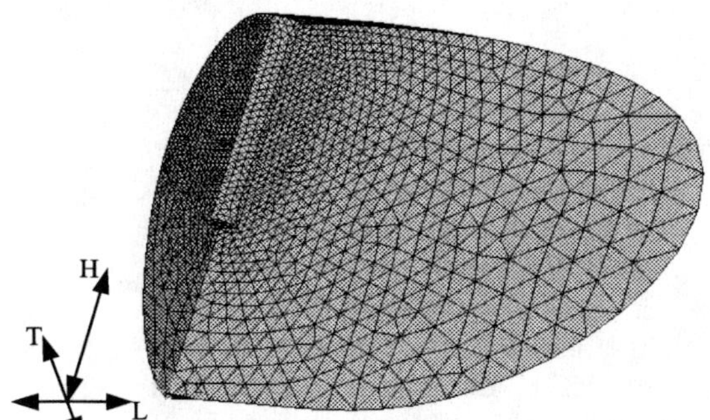

Fig. 3 The solid model with the finite element scheme. The orientations of length (L), height (H) and thickness (T) are given as well.

The domain was discretized with an effort to place a large number of elements near the microwave probe. The 3-D tetrahedral element was used in this study. The Pennes bioheat equation with the above-mentioned boundary conditions and FEM scheme was further solved using ANSYS (V5.7) to find the transient temperature distribution.

RESULTS

3.1 SAR measurements

The fitted constants are given as follows:

$A=1.24\times10^3$, n=0.9523, z_0=1.0515

a=-0.3065, b=2.783, c=-0.941, d=0.2295

The non-linear fitting results calculated from Eqn. 2 using the above constants are plotted in Fig. 4 with the average experimental results of four independent runs. The goodness of fit, R^2, is 0.95.

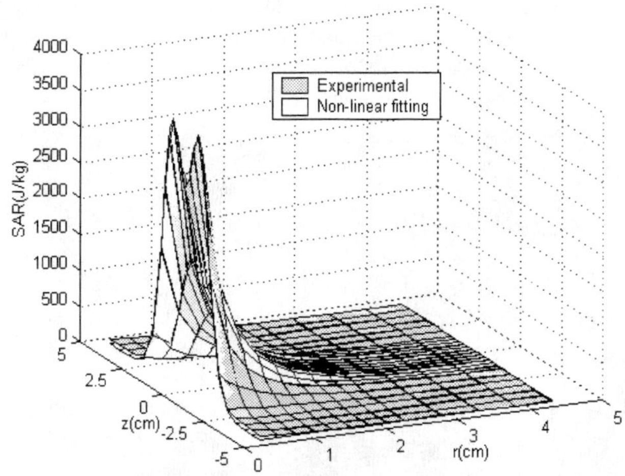

Fig. 4 Comparison between the experimental SAR and the SAR calculated by non-linear fitting.

3.2 Gross appearance of the kidney after therapy

Grossly, the ablative lesion in both *in vivo* and *in vitro* kidneys was characterized by a zone of dark, non-charred tissue immediately adjacent to the probe site with another zone of lighter tissue surrounding the dark zone. This color change is a typical indication of tissue injury due to heating. Representative figures of the gross appearance of kidneys after therapy are given in Fig. 5 for both *in vitro* and *in vivo* cases.

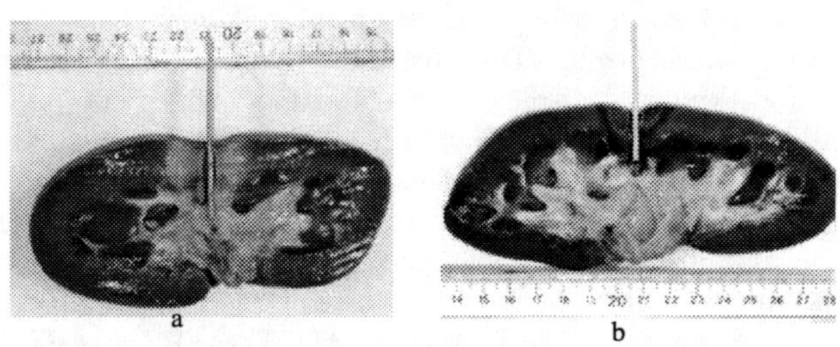

Fig. 5 Gross appearance of kidneys after therapy: a, *in vitro*, b, *in vivo*.

3.3 Histology

3.3.1 Injury at the cellular level

Following hyperthermic treatment and 48 hours tissue culture, H&E stained sections of both five *in vivo* and five *in vitro* porcine kidneys displayed tubular injury that was evident by two distinct zones distinguishable from unaffected renal parenchyma. The zone of injury immediately adjacent to the probe site was described as the complete ablative zone (CAZ) and distinguished by histological preservation of kidney architecture. The adjacent zone of injury, for descriptive purposes called the partial ablative zone (PAZ), was characterized by coagulative-type necrosis following tissue culture. The CAZ and PAZ are shown schematically in Fig. 2. Using criterions described in section 2.6, the sizes of these different injury zones were quantified and are summarized in Table 2. Representative histological appearances of the CAZ and PAZ are given in Fig. 6a and 6b.

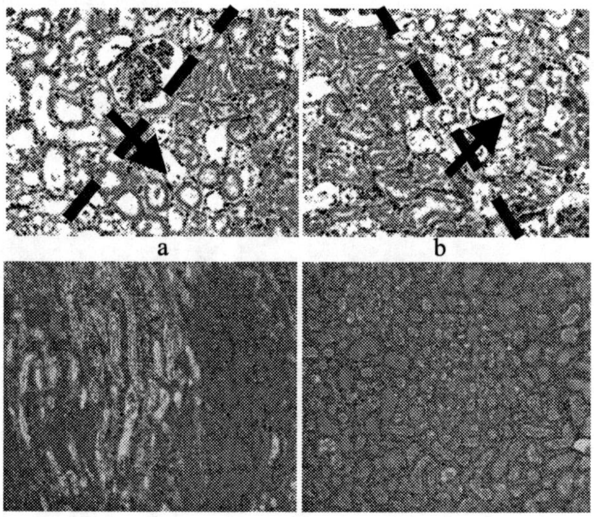

Fig. 6 Histological appearances at the cellular level (a and b, 400X) and the microvascular level (c and d, 200X): a, from complete ablative zone to partial ablative zone, b from partial ablative zone to normal zone, c, red zone, and d, pink zone.

3.3.2 Injury at the microvascular level

Following hyperthermic treatment and immediate tissue fixation in formalin, H&E stained sections of five *in vivo* porcine kidneys displayed two distinct zones of injury distinguishable from unaffected renal parenchyma. A zone of injury characterized by capillary congestion and red blood cell damage was described as the red zone (RZ). Another acute zone of injury was characterized by leakage of proteinacious fluid into the lumen of renal tubules. For descriptive purposes, this zone was called the pink zone (PZ). These RZ and PZ are also shown schematically in Fig. 2. Using criterions described in section 2.6, the sizes of these different injury zones were quantified and are summarized in Table 3. Representative histological appearances of RZ and PZ are given in Fig. 6c and 6d.

3.4 Thermal measurements

The measured thermal histories at different locations during therapy are shown in Fig. 7. The thermal histories *in vitro* (Fig. 7a) show a larger temperature elevation than *in vivo* (Fig. 7b) at each radial measurement location. Presumably, this was due to the cooling effect of blood perfusion *in vivo*. All of the *in vivo* and *in vitro* thermal histories at the probe tip show a sudden jump at 600s. This was due to shutdown of the coolant flow in the catheter.

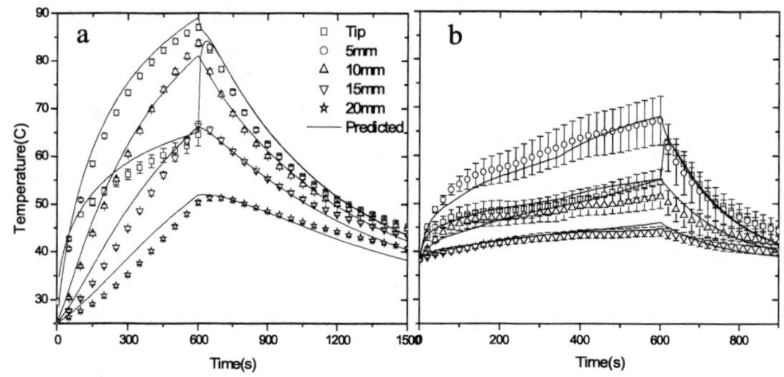

Fig. 7 Comparison between the predicted and measured thermal histories at several locations inside the kidney: a, *in vitro*, b, *in vivo*. The error bars represent standard error of the mean.

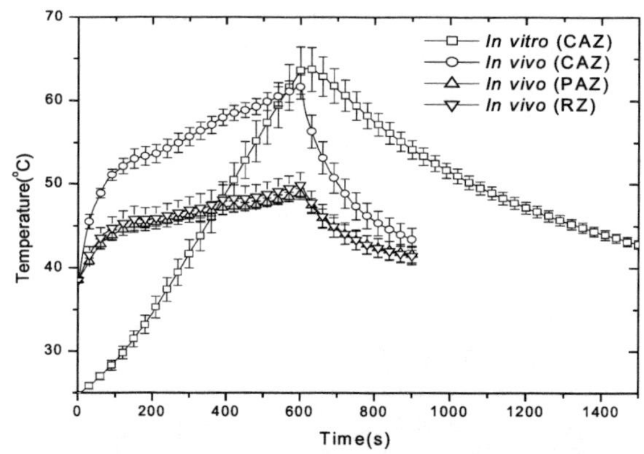

Fig. 8 Interpolated average thermal histories on the edge of various zones. The error bar represents the standard error of the mean.

3.5 Quantification of the kinetic parameters

3.5.1 Kinetic parameters for cellular injury

Using the thermal histories shown in Fig. 7, the thermal histories on the edge of different zones were determined by Lagrangian interpolation. The interpolated thermal histories on the edge of the completely ablated zone (CAZ) for five *in vitro* kidneys and their average value are given in Fig. 8. The average thermal history shows a peak value of 63.5°C. It also shows that a time period of 10 minutes above 50°C is required for complete ablation of the kidney tissue. The interpolated thermal histories on the edge of the complete ablative zone for five *in vivo* kidneys and their average value are also given in Fig. 8. The *in vivo* average thermal histories shows a peak temperature of 61°C with a time period of 10 minutes above 50°C.

Table 2 Summary of the results at the cellular level

	K	1	2	3	4	5	A±SEM
in vitro	Complete ablative zone						
	R	18.4	15.0	17.5	15.5	15.0	16.3 ±0.7
	F	0.46	0.14	0.22	0.13	0.09	0.2±0.07
	M	0.2	0.2	0.2	0.2	0.2	0.2±0.
	ΔH_c=98.7 kJ/mole ΔS_c=0.7 J/(mole·K)						
in vivo	Complete ablative zone						
	R	5.2	8.4	7.2	4.0	6.4	6.2 ± 0.8
	F	0.19	0.07	0.2	0.29	0.30	0.22± 0.04
	M	0.2	0.2	0.2	0.2	0.2	0.2±0.
	Partial ablative zone						
	R	10.0	14.0	12.3	10.4	15.0	12.3 ±1.0
	F	0.59	0.59	0.52	0.61	0.58	0.58±0.02
	M	0.6	0.6	0.6	0.6	0.6	0.6±0.
	ΔH_c=98.0 kJ/mole ΔS_c=0.3 J/(mole·K)						

Note: 'K' is kidney, 'R' is radius in mm calculated from the probe center, 'F' is fitted results, 'M' is measured results, 'A' is average, 'SEM' is standard error of the mean.

Using the procedures described in 2.8.2, the injury model was fit to the five *in vitro* thermal histories in Fig. 8 and their corresponding known injuries (i.e. 20% cellular viability). This process yielded an activation enthalpy of 98.7kJ/mole and activation entropy of 0.7J/(mole·K). Similarly, the fitting of *in vivo* thermal histories (PAZ and CAZ) in Fig. 8 and their corresponding injuries (20% or 60% cellular viability) gave an activation enthalpy of 98.0kJ/mole and an activation entropy of 0.3J/(mole·K). A summary of these results is given in Table 2 together with a comparison between the injury fit and actual injury.

3.5.2 Kinetic parameters for microvascular stasis

Similarly, the interpolated thermal histories on the edge of red zones for five *in vivo* kidneys are shown in Fig. 8. The average thermal histories show a peak temperature of 50°C. The overall thermal history was much lower than that on the edge of the complete ablative zone (CAZ). Similarly, the model described in section 2.9 was fit to *in vivo* thermal histories (RZ) in Fig. 8 and their corresponding injuries (20% tissue showing no microvascular stasis). This process yielded an activation enthalpy of 95.6kJ/mole and activation entropy of 1.2J/(mole·K). These results are summarized in Table 3.

Table 3 Summary of the results at the microvascular level

	K	1	2	3	4	5	A+SEM
in vivo	Red zone						
	R	10.9	13.5	10.9	7.0	9.8	10.4 ± 1.0
	F	0.29	0.21	0.09	0.11	0.33	0.21±0.05
	M	0.2	0.2	0.2	0.2	0.2	0.2±0.
	ΔH_s=95.6 kJ/mole ΔS_s=1.2 J/(mole·K)						

Note: 'K' is kidney, 'R' is radius in mm calculated from the probe center, 'F' is fitted results, 'M' is measured results, 'A' is average, 'SEM' is the standard error of the mean.

3.6 Characterization of blood perfusion

The blood perfusion was found indirectly by matching the predicted and measured *in vivo* thermal histories. In order to make blood perfusion the only unknown parameter, the FEM model was first verified by the measured *in vitro* thermal histories. Using the input parameters given in section 2.10, the predicted *in vitro* thermal histories match the experimental ones very well as shown in Fig. 7a. Assuming the intrinsic thermal properties of the kidney tissue and the heat transfer coefficient are the same for both *in vivo* and *in vitro* cases, the only unknown left for *in vivo* thermal prediction is the blood perfusion term. By using the blood flow model described by Eqn. 9 and by trial and error, a relationship between F(DS) and DS was found and is shown in Fig. 9. When input into the *in vivo* thermal model, a good match between the predicted and measured thermal histories was obtained (see Fig. 7b).

As can be seen from Fig. 9, the blood perfusion is higher than the baseline level until the degree of stasis reaches approximately 20%. There exists a plateau when the degree of stasis is between 40-60%, and then there is some residual blood perfusion even when the degree of stasis from histology is judged to be 100%.

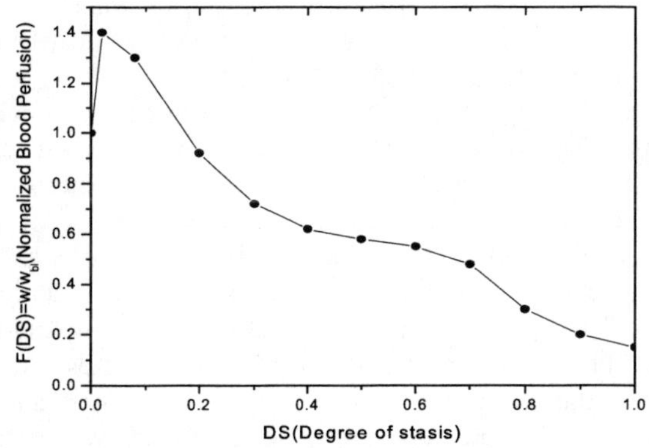

Fig. 9 Relationship between the normalized blood perfusion used in the thermal model and the degree of vascular stasis predicted using the kinetical parameters from microvascular stasis analysis.

DISCUSSION

4.1 Injury measurements

At the cellular level, injury measurements from histology showed that the average complete ablative zone (CAZ) of the *in vitro* cases was 10mm larger than that of the *in vivo* cases (16.3mm vs. 6.2mm). Presumably, this difference was caused by the cooling effect of blood perfusion *in vivo*, considering that the kidney is a highly perfused organ. In addition, using capillary congestion and red blood cell damage as the injury criterion, one can define a red zone (Fig. 2) that is much larger than the cellular complete ablative zone (10.4mm vs. 6.2mm). Since the red zone represents microvascular stasis, one can expect that, given time, part of the cells within this zone will succumb due to the lack of oxygen and nutrients. A former study [41] showed a strong correlation between microvascular stasis and tissue coagulative necrosis for both normal and tumor tissue after 1, 3, and 7 days post thermal injury. In the current study, kidneys remained within the animal for about 2 hours prior to tissue harvest and fixation or two days culture. Two hours is too short for ischemic necrosis to develop histologically. A pink zone (15.6±1.3mm) extending beyond the red zone was characterized by leakage of proteinacious fluid into the lumen of renal tubules. However, the edge of this zone was very irregular (data not shown). Further, some clear microvascular injury was identified outside the pink zone by the purple color of the capillary wall (due to the denaturation of collagen, data not shown). These phenomena may be due to the outward movement of hot blood from the near field surrounding the probe, which we term "thermal tracking".

4.2 Thermal prediction and blood perfusion

The error bars of the *in vitro* thermal histories are very small (Fig. 7a). This indicates the probe placement was quite accurate. The error bars of the *in vivo* thermal histories are larger. This probably was due to the variation of baseline blood perfusion from individual to individual as evidenced by our measured blood perfusion, which ranged from 0.5ml/(mg·min) to 2.2ml/(mg·min) with an average of 1.2 ± 0.5 ml/(mg·min). However, the measurement technique may introduce error as well. The predicted *in vitro* thermal histories matched the measured thermal histories very well, which indicates the thermal properties and relevant input parameters were appropriate. The predicted *in vivo* thermal histories reasonably capture the average measured thermal histories of the five kidneys studied. The use of the average thermal history was motivated by the belief that the blood flow response of the kidney to thermal insult would be best studied as an average response in the kidneys. This response in the present study is given in Fig 9. The nonlinear relationship between the normalized blood flow and the degree of stasis indicates that the dynamics of the change of blood flow in kidney tissue due to thermal exposure could not simply be described by the first order rate process. The initial increase in blood perfusion is due to physiological response of the normal tissue to compensate for the abnormal supraphysiological temperature. Similar phenomena have been observed in several other studies on normal tissue, while for tumors, many studies showed a drop of blood perfusion without an initial increase [26, 27]. The plateau and the existence of residual blood perfusion in Fig. 9 are hard to explain, but may be due to the complex interaction between hydrostatic pressure, change of blood rheologic properties and microvascular injury [37].

4.3 Correlation between the thermal and injury measurements

As shown in Fig 8, the average peak temperature at the edge of complete ablative zones *in vitro* (63.5°C) is close to that of *in vivo* (61°C). These peak temperatures are lower than that of 84°C in studies by Kigure et al [42-43] when investigating microwave thermal therapy of VX-2 rat renal tumor. Both the average *in vitro* and *in vivo* thermal history on the border of complete ablative zone show a thermal history above 50°C for 10 minutes, while in Kigure's study, only approximately two minutes above 50°C was shown. The longer period in our thermal histories probably compensates for the lower peak temperature when comparing to Kigure's results. This comparison indicates that injury is dependent on both temperature and time. Characterizing injury only by a critical temperature is thus misleading. When fitting the kinetic model to relevant data, the thermal histories in Fig. 8 yielded very similar activation enthalpy and entropy as shown in Table 2. This suggests that, for acute tissue thermal injury, the injury mechanisms at the cellular level were similar for both *in vivo* and *in vitro*. The *in vivo* average critical thermal history at the edge of the red zone is close to that of the PAZ, but very different from that of the CAZ. This indicates that the threshold of microvascular injury is lower than that of cellular injury in tissue. This is further evidenced by the smaller activation enthalpy as given in Table 2 and 3, although the difference is not large, as also shown in another study [41]. The activation enthalpy found in this study is much smaller than those found from studies of cultured single cells (200-1600kJ/mole)[36], but it is close to those of several tissue studies (20-200kJ/mole) [44-45] and overall protein denaturation in cultured cells (around 100kJ/mole) [46]. This suggests thermal injury of tissues might correlate to overall protein denaturation in cells regardless of whether they are the constitutive cells of the tissue or endothelial cells lining the capillary wall.

SUMMARY

The thermal and injury behavior during microwave thermal therapy of normal porcine kidney have been systematically studied in this paper. At the cellular level, for both *in vitro* and *in vivo* kidneys, histological evaluation showed a complete ablative zone characterized by preservation of kidney architecture and a partially ablated zone extending beyond the complete ablative zone characterized by coagulative-type necrosis. At the microvascular level, *in vivo* histology showed a red zone characterized by capillary congestion and red blood cell damage and a pink zone characterized by leakage of proteinacious fluid into the lumen of renal tubules. The average radius of the complete ablative zone *in vitro* was 16.3 ± 0.7mm which was 10mm larger than that of 6.2 ± 0.8mm for the *in vivo* case. The critical thermal histories at these two locations showed peak temperatures of 63.5°C *in vitro* and 61°C *in vivo*. These data also indicated a thermal history above 50°C for 10 minutes was required to reach complete ablation at the cellular level in this study. A more general way to evaluate the thermal histories is to fit them with a kinetic model. By doing so, an activation enthalpy around 98 kJ/mole was found for both *in vivo* and *in vitro* cases, indicating similar injury mechanisms at the cellular level. Measurement of the red zone gave an average radius of 10.4 ± 1 mm, which was 4mm larger than that of the complete ablative zone *in vivo*. Since cells would not be able to survive without a sufficient supply of oxygen and nutrition, we expected the complete ablative zone would be larger than 6.2mm if kidneys were left within animals for a longer time.

The 3-D FEM thermal model was validated by comparison to the *in vitro* thermal measurements. This model was then applied to the *in vivo* case. By adjusting the blood perfusion, the predicted thermal histories *in vivo* matched the measured ones well. Blood perfusion was found to be temperature and time dependent and was correlated to the degree of microvascular stasis. Blood perfusion was characterized by an initial increase followed quickly by a drop with a small plateau likely due to complex rheological reaction of blood vessels due to heating. This exercise suggests that the first order rate model is perhaps insufficient to model the blood perfusion response in kidney due to localized microwave thermal therapy.

In conclusion, our acute study of interstitial microwave thermal therapy of porcine kidneys shows that thermal and injury behavior involved in this technique are predictable and can be used to ablate small diseased renal masses. Further chronic studies to investigate the host-mediated effect, in particular, a comparison of red zone to ultimate necrotic lesion *in vivo,* are warranted.

ACKNOWLEDGMENTS

Thanks to Urologix Inc. (Polymouth, MN) for their support of this study and the Supercomputing Institute at the University of Minnesota for a grant covering computing time.

REFERENCES

[1] Chow, W. H., Devesa, S. S., Warren, J. L. and Fraumeni, J. F. Jr., "Rising incidence of renal cell cancer in the United States", *JAMA*, Vol. 281, 1999, pp. 1628-1631

[2] Reddan, D. N., Ganesh, G. V., and Polscik, T. J., "Management of small renal tumors: an Overview", The American Journal of Medicine, Vol. 10, 2001, pp. 558-562

[3] Murphy, D. P. and Gill, I. S., " Energy-based renal tumor ablation: a review", Seminar in Urologic Oncology, Vol. 19, No. 2, 2001, pp. 133-140

[4] Savage, S. J. and Gill I. S., " Renal tumor ablation: energy-based technologies", World J. Urol., Vol. 18, 2000, pp283-288

[5] Nakada, S. Y., "Present state of the art: Newer techniques for surgical ablation of renal neoplasms", Contemporary Urology, Oct., 1999, pp68-80

[6] King, R. W. P., Trembly, B. S., and Strohbehn, J. W., "The electromagnetic field of an insulated antenna in a conducting or dielectric medium", IEEE Trans. on Microwave Theory and Tech., Vol. 31, No. 7, 1983, pp574-583

[7] Casey, J. P. and Bansal, R., "The near field of an insulated dipole in a dissipative dielectric medium", IEEE Trans. on Microwave theory and Tech., Vol. 34, No. 4, 1986, pp459-463

[8] Gentili, G. B., Gori, F., and Leoncini, M., " Electromagnetic and thermal models of a water cooled dipole radiating in biological tissue", IEEE Trans. Biomedical Eng., Vol. 38, No. 1, 1991, pp98-102

[9] Wu, L., and Su, D. W., "Design and analysis of an asymmetrically fed insulated coaxial slot antenna with enhanced tip-heating performance", IEEE Trans. on Microwave Theory and Tech., Vol. 48, No. 8, 2000, pp1369-1374

[10] Zhu, L., Xu, L. X., and Chencinski, N., "Quantification of the 3-D electromagnetic power Absorption rate in tissue during transurethral prostatic microwave thermotherapy using heat transfer model", IEEE Trans. on Biomedical Engineering, Vol. 45, NO. 9, 1998, pp1163-1172

[11] Sherar, M. D., Gladman, A. S., Davidson, S. R. H., Trachtenberg, J., and Gertner, M. R., "Helical antenna arrays for interstitial microwave thermal therapy for prostate cancer: tissue phantom testing and simulations for treatment", Phys. Med. Biol., Vol. 46, 2001, pp. 1905-1918

[12] Ryan, T. P., "Comparison of six microwave antennae for hyperthermia treatment of cancer: SAR results for single antennae and arrays", Int. J. Radiat. Oncol. Biol. Phys., Vol. 21, 1991, pp403-413

[13] Satoh, T., Stauffer, P. R., and Fike, J. R., " Thermal distribution studies of helical coil microwave antennae for interstitial hyperthermia", Int. J. Radiat. Oncol. Biol. Phys., Vol. 15, 1988, pp1209-1218

[14] Trembly, B. S., Douple, E. B., Ryan, T. P., and Hoopes, P. J., " Effect of phase modulation on the temperature distribution of a microwave hyperthermia antenna array *in vivo*", Int. J. Hyperthermia, Vol. 10, No. 5, 1995, pp691-705

[15] Roos, D., Seegenschmied, M. H., Klautke, G., Erb, J., and Sorbe, B., "A new microwave applicator with integrated cooling system for intracavitary hyperthermia of vaginal carcinoma", Int. J. Hyperthermia, Vol. 12, No. 6, 1996, pp743-756

[16] Debicki, P. S., Okoniewski, M., Okoniewska, E, Shrivastava, P. N., Debicka, A. M., Baert, L. V. and Petrovich, Z., "Cooled microwave transrectal applicator with adjustable directional beam for prostate treatment", Int. J. Hyperthermia, Vol. 11, No. 1, 1995, pp95-108

[17] Martin, G. T., Haddad, M. G., Cravalho, E. G., and Bowman, H. F., "Thermal model for the local microwave hyperthermia treatment of benign prostatic hyperthermia", IEEE Trans. on Biomedical Eng., Vol. 39, No. 8, 1992, pp836-843

[18] Xu, L. X., Rudie, E., and Holmes, K. R., "Transurethral thermal therapy (T3) for the treatment of benign prostatic hyperplasia (BPH) in the canine: analysis using Pennes bioheat equation", Adv. Bioheat Mass Transfer, HTD-vol. 268, 1993, pp31-35

[19] Yuan, D. Y., Valvano, J. W., Rudie, E. N., and Xu, L. X., "2-D finite difference modeling of microwave heating in the prostate", Adv. Biol. Heat, Mass Transfer in Biotech., Vol. HTD-322, 1995, pp107-115

[20] Liang, P., Dong, B., Yu, X., Cheng, Z., Su, L., Peng, J., Nan, Q., and Wang, H., "Computer-aided dynamic simulation of microwave-induced thermal distribution in coagulation of liver cancer", IEEE Trans. on Biomed. Engineering, Vol. 48, No. 7, 2001, pp. 821-829

[21] Sherar, M. D., Moriarty, J. A., Kolios, M. C., Chen, J. C., Peters, R. D., Ang, L. C., Henkelman, R. S., Bronskill, M. J., and Kucharcyk, W., "Comparison of thermal damage calculated using magnetic resonance thermometry, with magnetic resonance imaging post-treatment and histology, after interstitial microwave thermal therapy of rabbit brain", Phys. Med. Biol., Vol. 45, 2000, pp3563-3576

[22] Skinner, M. G., Iizuka, M. N., Kolios, M. C., and Sherar, M. D., "A theoretical comparison of energy sources-microwave, ultrasound and laser-for interstitial thermal therapy", Phys. Med. Biol., Vol. 43, 1998, pp3535-3547

[23] Pennes, H. H., "Analysis of tissue and arterial blood temperatures in the resting forearm", J. Applied Physiology, Vol. 1, 1948, pp93-122

[24] Arkin, H., Xu, L. X., and Holmes, K. R., "Recent developments in modeling heat transfer in blood perfused tissues", IEEE Trans. Biomed. Eng., Vol. 41, 1994, pp97-107

[25] Xu, L. X., Zhu, L., and Holmes, K. R., "Blood perfusion measurements in the canine prostate during transurethral hyperthermia", Annals of the New York Academy of Sciences, Vol. 858, 1998, pp21-29

[26] Vaupel, P., "Pathophysiological mechanisms of hyperthermia in cancer therapy", in *Biological basis of oncologic thermotherapy*, Ed: Gautherie, M., Springer-Verlag, Berlin, Heidelberg, 1990, pp73-134

[27] Song, C. W., "Effect of local blood hyperthermia on blood flow and microenvironment", Cancer Res. [Suppl], Vol. 44, 1984, pp4721s-4730s

[28] Brown, S. L., Hunt, J. W., and Hill, R. P., "Differential thermal sensitivity of tumor and normal tissue microvascular during hyperthermia", Int. J. Hyperthermia, Vol. 8, NO. 4, 1992, pp501-514

[29] Dudar, T. E., and Jain, R. K., "Differential response of normal and tumor microcirculation to hyperthermia", Cancer Res., Vol. 44, 1984, pp605-612

[30] Diller, K. R., Valvano, J. W., and Pearce, J. A., " Bioheat transfer", Chapter 4.4 in *CRC Thermal Engineering Handbook*, Ed: Kreith, F., 2000, pp4-114-187

[31] Bhowmick, S., "Supraphysiological thermal therapy in the treatment of prostate cancer: linking thermal history to cellular and tissue level injury mechanisms", *Ph. D. thesis, University of Minnesot*a, 2000

[32] Pearce, J. and Thomsen, S., "Rate process analysis of thermal damage", in *Optical Thermal Response of Laser-Irradiated Tissue*, Eds: A. J. Welch and M. J. C. van Germert, Plenum Press, New York, 1995, pp561-606

[33] Cravalho, E. G., Toner, M., Gaylor, D. C. and Lee, R. C., "Response of cells to supraphysiological temperatures: experimental measurements and kinetic models", Chapter in *Electrical Trauma: The Pathophysiology, Manifestations and Clinical Management.* Ed: Lee, R. C., Cravalho, E. G., Burke, J. F., Cambridge University Press, 1992

[34] Sapareto, S. A., and Dewey W. C., "Thermal dose determination in cancer therapy", Int. J. Radiat. Oncol. Biol. Phys., Vol. 10, 1984, pp787-800

[35] Tropea, B. I. and Lee, R. C., "Thermal injury kinetics in electrical trauma", J. Biomech. Eng., Vol. 114, 1992, pp241-250

[36] Andreuccetti, D., Bini, M., Ignesti, A.,Olmi, R., Rubina, N., and Vanni, R., "Use of polyacrylanide as a tissue-equivalent material in the microwave range", IEEE Trans. Biomed. Eng., Vol. 35, 1988, pp275-277

[37] Lund, T., Onarheim, H., and Reed, R. K., "Pathogenesis of edema formation in burn injuries", World J. of Surg., Vol. 16, 1992, pp2-9

[38] Himmelblau, D. M., "Applied nonlinear programming", MCGraw- Hill Inc., 1972

[39] Nelder, J. A., and Mead, R., "A simplex method for function minimization", Computer Journal, Vol. 7, 1964, pp308-313

[40] Spendly, W., Hext, G. R., and Himsworth, F. R., "Sequential application of simplex designs in optimization and evolutionary operation", Technometrics, Vol. 4, 1962, p441

[41] Bhowmick, S., Hoffman, N. E., and Bischof, J. C., "Thermal therapy of prostate tumor tissue in the dorsal skin flap chamber", Microvascular Research, Vol. 64, pp170-173, 2002

[42] Kigure, T., Harada, T., Turi, Y., and Satoh, Y., "Ultrasound –guided microwave thermotherapy on a VX-2 carcinoma implanted in Rabbit kidney", Ultrasound in Med. & Biol., Vol. 21, No. 5, 1995, pp649-655

[43] Kigure, T., Harada, T., Turi, Y., Satoh, Y., and Yoshida, K., "Laparoscopic microwave thermotherapy on small renal tumors: experimental studies using implanted VX-2 tumors in rabbits", European Urology, Vol. 30, 1996, pp377-382

[44] Bhowmick, S., Pedersen, J., and Bischof, J. C., "Investigation of corneal/scleral burning during cataract surgery", Advans. Heat Mass Transfer in Biotech., HTD-Vol. 368/BED-Vol. 47, 2000, pp73-74

[45] Bhowmick, S., Bhowmick, P., Coad, J. E., and Bischof, J. C., " *In vitro* assessment of the efficacy of thermal therapy in human prostatic hyperplasia tissue", Fundamentals of Heat Transfer, IMECE 2001/ HTD-24434 (CD version)

[46] He, X., Wolkers, W., Crowe, J., and Bischof, J., "Investigation of *in situ* thermal denaturation of proteins in Dunning AT-1 rat prostate cancer cells using DSC and FTIR", to be submitted to IMECE 2002

Proceedings of IMECE2002
ASME International Mechanical Engineering Congress & Exposition
November 17–22, 2002, New Orleans, Louisiana

IMECE2002-32055

EXPERIMENTAL AND NUMERICAL STUDIES OF HEAT TRANSFER IN HUMAN UTERI DURING CRYOABLATION

Jim S. Chen and Kevin Agnissey
Department of Mechanical Engineering, College of Engineering

Marla Wolfson, Charles Philips, and Thomas Shaffer
Department of Physiology, School of Medicine

Temple University, Philadelphia, PA 19122

ABSTRACT

This paper presents experimental and numerical studies of transient heat transfer inside the uterus during application of a PFC (perfluorochemical) fluid into the endometrium cavity in order to achieve cryoablation. The numerical prediction is based on a 1-D finite difference method of the bio-heat equation using the Crank Nicolson scheme. The numerical method is first validated by a 1-D physical model by measuring temperature history at several locations within a silicone rubber sheet. Good agreement, thus positive predictability, was obtained by comparing numerical predictions with the experimental data obtained from eight intact, hysterectomized uteri during cryoablation.

INTRODUCTION

Menorrhagia, defined as excessively heavy or prolonged bleeding, affects approximately twenty percent of women worldwide (Cooper and Erikson, 2000). This abnormal uterine bleeding may be caused directly by uterine abnormalities or hormone imbalance. Other types of abnormal uterine bleeding are too-frequent menstrual periods, bleeding between periods, and bleeding after menopause (Best and Taylor, 1991). In a normal menstrual cycle, there is a balance between estrogen and progesterone, two hormones in the body. These hormones regulate the buildup of the endometrium (uterine lining of blood and tissue), which is shed each month during menstruation. Menorrhagia can occur because of an imbalance between estrogen and progesterone. As a result of the imbalance, the endometrium keeps building up. When it is eventually shed, there is heavy bleeding. Because hormone imbalances are often present in adolescents and in women approaching menopause, this type of menorrhagia known as dysfunctional uterine bleeding is fairly common in these groups (DeCherney et al., 1987). Another frequent cause of Menorrhagia is uterine fibroids growth. Together, hormone imbalance and fibroids account for about 80% of menorrhagia cases. Other causes include endometrial cancer, inflammation or infection of the vagina, cervix, or pelvic organs, polyps (small growths on the cervical or uterine wall), thyroid conditions, and liver, kidney or blood diseases, and the use of blood-thinning drugs. Although cancer is not a frequent cause of menorrhagia, it should not be ruled out in women > 35 years of age or in younger women with longer duration of exposure to unopposed estrogen.

Localized cryoablation techniques have been developed to treat endometrial dysfunction but with little success (Cooper and Erikson, 2000). The present research introduces a new technique using a PFC (perfluorochemical) fluid that allows cryoablation to be achieved on the entire inner lineage of the endometrium, including the cornua and other anatomic recesses. PFCs are synthetic fluorinated hydrocarbons, formulated by substituting fluorine for hydrogen atoms on the common organic compounds. They are clear, nontoxic, insoluble in water and biochemically inert. The low surface tension of PFCs in the pour point range allows for uniform contact with the uterine cavity, supporting the possibility for uniform ablation both geometrically and in depth of freeze. In addition, PFCs are not metabolized and their elimination by the body is by evaporation during exhalation or transpiration through the skin of the intact molecule (Shaffer et al., 1998).

Prior studies have shown that a rapid freeze followed by a slow thaw results in maximum damage of tissues during a cryoablation (Song and Jiji, 1988). Freezing problems during cryosurgery have been investigated by researchers including Tatsutani and Rubinsky (1998) who studied intracellular ice nucleation and its effect on prostatic cancel cell destruction, Rabin and Shitzer (1998) who numerically solved the multidimensional freezing problem based on a cylindrical cryoprobe, and Hoffmann and Bischof (2001) who examined thermal response during cryosurgery of normal and tumor tissue in the dorsal skin flap chamber.

In this study, it is estimated that cryoablation can be effectively performed by freezing tissues of the endometrium with a thickness of 3-5 mm at -30°C for a period of 200 seconds. The objective of this paper is to develop a numerical model capable of predicting the temperature distribution in the uterus including both the endometrium and the myometrium, so as to control the desired depth of tissue necrosis in a uterus during an endometrial cryoablation.

NOMENCLATURE

c_p	specific heat [J/kg °C]
q	heat flux [W/m^2]
t	time variable
T	temperature variable
T_0	initial temperature [°C]
$T_\infty(t)$	fluid temperature [°C]
$T_s(t)$	cold (endometrium) surface temperature [°C]
$T_L(t)$	warm surface or body temperature [°C]
Δt	time increment and sampling interval [s]
Δx	x-increment [m]
k	thermal conductivity [W/m K]
L	thickness of the sample [m]
ρ	density [kg/m^3]
α	thermal diffusivity [m^2/s]

NUMERICAL MODEL

To theoretically evaluate the transient temperature distribution in the uteri, a one-dimensional thermal model is developed in the Cartesian coordinates. Several assumptions are made to simply the bioheat transfer equation (Weinbaum and Jiji, 1985; Shitzer and Eberhart, 1985; Hartnett et al., 1992): a) invariance of the thermophysical properties, b) negligible blood perfusion and internal heat generation, c) thermal continuity between the endometrium and the myometrium, and d)) negligible latent due to phase change.

With the above assumptions, the governing equation that needs to be solved is the heat conduction equation with appropriate initial and boundary conditions:

$$\frac{\partial T(x,t)}{\partial t} = \alpha \frac{\partial^2 T(x,t)}{\partial x^2} \qquad (1)$$

where T is the temperature and α is the thermal diffusivity of the endometrium. A finite difference method based on the Crank-Nicolson algorithm is used because of its numerical stability and accuracy (Ozisik, 1994).

Two types of boundary conditions on the cold surface (endometrium surface) are considered in the present study. The 1st type is based on the measured surface temperature history, which can be easily obtained in the laboratory but not so easily during cryosurgery. The 2nd type is based on the measured PFC temperature along with the convective heat transfer coefficient, both of which are time dependent. While the PFC temperatures can be recorded, the convective heat transfer coefficient must be predicted based on laboratory experiments simulating the actual cryoablation procedure. The boundary condition inside the uteri (inside the myometrium) is considered to be in equilibrium with body temperature, i.e., T = 37°C. In cryosurgery, this assumption is only accurate when the blood perfusion and metabolic heat generation is sufficient to maintain the body temperature inside the uterus or when the uterus is sufficiently thick.

The finite difference method (FDM) based on the Crank-Nicolson method scheme is used, because its accuracy and unconditional stability. It is accurate to the second order in both time and space, i.e., the truncation error is on the order of $(\Delta t)^3$ as opposed to $(\Delta t)^2$ in the simple explicit and implicit methods. The program was coded in dimensionless forms to minimize round off errors. The accuracy of the 1-D FDM is first verified by comparing the numerical solution with the exact solution for a 20 mm thick plate initially at 27°C exposed to a sudden temperature change (-80°C) at the surface at x=0, t ≥0. In this case study, α = 1.8 x 10^{-7}m^2/s, Δx = 1mm and Δt of 10 seconds are chosen to match the thermal diffusivity, depth tolerance and sampling interval in the *in vitro* uteri experiments. The numerical solutions are almost identical to the exact solution with errors less than 0.1°C.

EXPERIMENTAL SETUP

Before the implementation of the 1-D model on the biological model, the FDM model was first validated using a 1-D physical model. The physical model was based on a silicon rubber sheet with known thermophysical properties. The silicon rubber was chosen because its texture thermal diffusivity (α = 0.7 x 10^{-7}m^2/s) is similar to human uteri tissues.

1-D Silicon Rubber Experiment

As shown in Fig. 1, a 1-D physical model was designed and built using a silicon rubber sheet with α = 0.7x10^{-7}m^2/s. The sample is flat, 10 cm long, 10 cm wide and 1.2 cm thick. Warm water maintained at 37°C is used to circulate one side of the sheet. The other side is cooled at a temperature below -60°C. Two sets of experiments are performed with this apparatus. The first experiment uses a mixture of dry ice and acetone as the cooling fluid. This experiment was designed to validate the numerical simulation. The second experiment uses

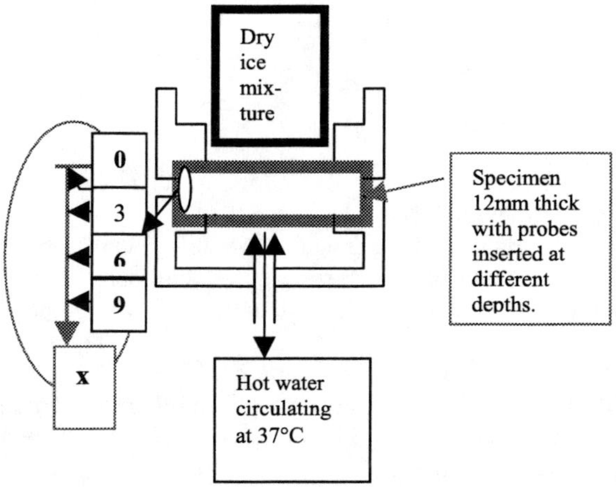

Fig. 1 Experimental apparatus for the silicone rubber sheet.

a mixture of dry ice and PFC for simulating the cryoablation process. Dry ice is chosen to maintain the fluid temperature constant during the cooling process. Several type J Cole thermocouple probes are inserted at different depth in the 12.5 mm thick silicone rubber sheet. One probe is on the cold side, a second at 3.125 mm, a third at 6.25 mm, a fourth at 9.375 mm, and a fifth on the external surface (12.5 mm) of the sheet on the warm side. Another thermocouple is used to measure the cold fluid mixture temperature. The thermocouple probes were connected to a Cole-Parmer 12 Channel Scanning Thermocouple Thermometer (model # 92000-00 benchtop) with a temperature range of –200 °C to 1000°C, a resolution of 0.1°C, and an accuracy of 0.5°C. The multi-channel data acquisition system was equipped with cold junction compensation and signal amplification. The thermocouple probes with fine tips have a time constant about 1.0 s. The temperatures are recorded every 4 or 10 s.

Biological *in vitro* Experiment

Uteri were obtained from 8 patients who were undergoing uterotomies for dysfunctional uterine bleeding. Access to this tissue was granted through Temple University Institutional Review Board. Each removed uterus was placed in an isotonic saline bath, as shown in Fig. 2. The bath was maintained at 37°C with a heater circulator. Two parallel tubes, with the inflow and outflow diameter of respectively 0.4 and 0.3 cm were inserted in the uterus through the cervix. Four 4-inch long Cole Parmer thermocouple probes with a time constant of 1.0 second in water are placed at different locations in the endometrium and the myometrium. One was placed in the superficial myometrium, one each at the right and left lateral walls and one deep in the myometrial wall that was referenced as the anterior probe. In addition, three other thermocouples were installed to measure the temperature of the inflow fluid, the temperature of the outflow fluid, and the temperature of the saline bath.

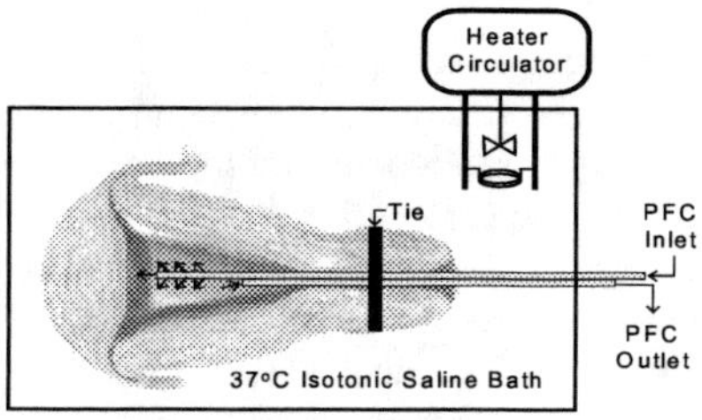

Fig. 2 Schematic of the temperature bath apparatus.

Cold PFC fluid was infused through the uterus and the temperature of each probe was recorded. The probes were inserted, through the exterior surface of the uterus, to different depths of the uterine muscle. The exact positions of the probes were measured with calibrated calipers at the end of the procedure when the uterus was dissected longitudinally. Pre-cooled (-80°C) PFC liquid was loaded into 60 ml syringes. Immediately, the head of the syringe was locked into the inlet tube 1 and the fluid was injected into the uterine cavity. The properties of the PFC 216 used in the experiment and water at 20°C are listed in Table 1.

The PFC fluid flowed out of the uterine cavity through the outlet tube 2 to a waste container. As each syringe emptied into the uterine cavity, another syringe was loaded, locked into tube 1 and the process was repeated 6 times over a time period of 400 seconds. The average flow rate was 0.5 – 1.0 ml/sec. The temperature was read every 10 seconds on each probe using the same data acquisition system described in the previous section.

RESULTS AND DISCUSSION

Silicon Rubber

Fig. 3 shows the numerical and measured temperatures at various distances from the dry ice and acetone mixture for the duration of 400 seconds. The bottom curve is the measured

Table 1 Properties of PFC 216

Properties	PFC	Water at 20°C
Molecular Weight	902	
Pour Point, ,°C	-108	0
Density, g/cm^3	1.84 (20°C)	1.0
Density, g/cm^3	1.93(-20°C)	1.0
Kinematic Viscosity, cs	4.8 (20°C)	1.0
Kinematic Viscosity, cs	647 (50°C)	1.0
Vapor pressure, torr	<1 (25°C)	18
Specific Heat, Btu/lbm-°F	0.26 (25°C)	1.0
Surface Tension, dynes/cm	15.4 (25°C)	70

fluid temperature, $T_\infty(t)$, the curve slightly above is the measured cold surface temperature, $T_s(t)$, and the top curve is the measured hot side temperature, $T_L(t)$. The numerical results are based on the 1-D Crank-Nicolson method with measured boundary conditions, $T_s(t)$ and $T_L(t)$, for the cold surface (x = 0) and the warm surface (x = L) temperatures, respectively. Because the silicone rubber is thin (thickness, L = 12.5 mm), the warm side temperature $T_L(t)$ decreases from 37°C to 29°C. The assumption of constant temperature at x = L is only valid for thicker samples. Strong agreement between the numerical results and the measured temperatures are obtained at x = 3.125, 6.35, and 9.375 mm. This further validates the 1-D FDM model. At t = 400 s, the fluid and cold surface temperatures approach equilibrium values while the internal temperatures continue to decrease due to heat conduction.

After the internal surface temperatures are determined, the transient surface heat flux (q'') and convective heat transfer coefficient (h) values are numerically calculated using

$$q'' = -k\frac{\partial T}{\partial x}\bigg|_{x=0} \qquad (2)$$

$$h(t) = \frac{-k\partial T/\partial x\big|_{x=0}}{[T_\infty(t) - T_s(t)]} \qquad (3)$$

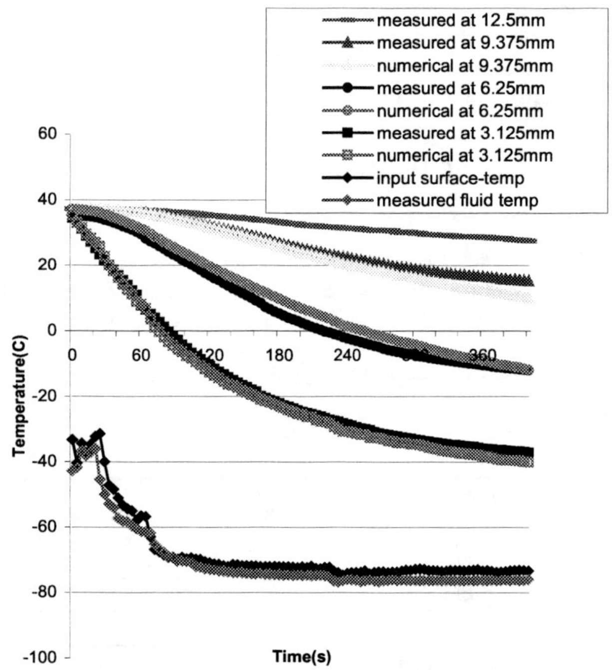

Fig. 3 Temperature distribution inside a sheet of silicone rubber cooled with dry ice and acetone (Δx=0.3125mm, Δt=4s).

The finite difference form of Eq. (3) with a second order

$$h = \frac{-k(-3T_s + 4T_1 - T_2)}{2\Delta x(T_\infty - T_s)} \qquad (4)$$

In our numerical analysis, k=0.2W/mK for the silicone rubber, Δx = 0.3125 mm, and Δt = 4 s are used.

Fig. 4 shows the heat transfer coefficient versus time for the dry ice-acetone mixture experiment. The data for t = 0 – 100 s are not shown due to mishandling of the temperature probes in the beginning of the pouring process. The h values decrease from an initial peak value of 4,500 W/m^{2}K to 800 W/m^{2}K at the end of the experiment. The higher initial value is attributed to the initial fluid motion (forced convection) and the larger temperature difference (free convection).

Because our experiment was performed with a cold fluid that was almost stationary, the calculated h values are less than those (h = 4,000 W/m$^{2\cdot\circ}$C) obtained by Tunnel et al. (2002) who used R-134a in a Cryogen spray cooling process. They solved the inverse heat conduction (IHC) problems using measured internal temperature measurements. In their study, the heat fluxes decreased with time, but the heat transfer coefficients were relatively constant due to dominating forced convection.

Fig. 5 shows the heat fluxes for two experiments with the silicone rubber sheet cooled with dry ice and acetone, one experiment at 4 s (the lower) and the other at 10 s (the upper curve) sampling time intervals. We noticed that the two curves

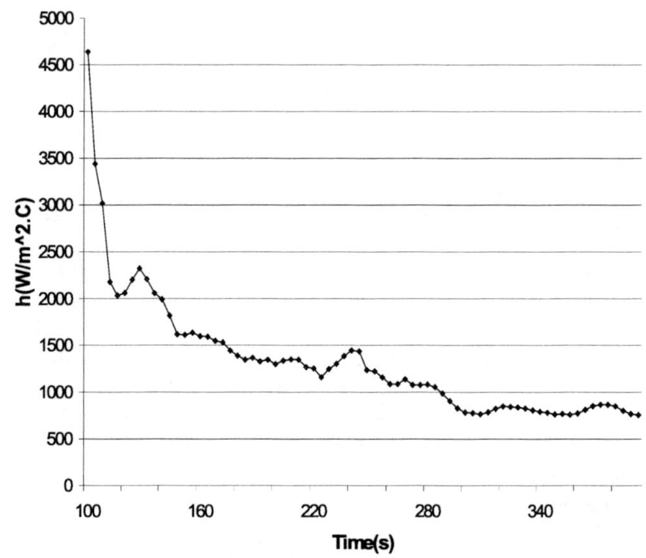

Fig. 4 Heat transfer coefficients h(t) for experiment with silicone rubber, dry ice and acetone.

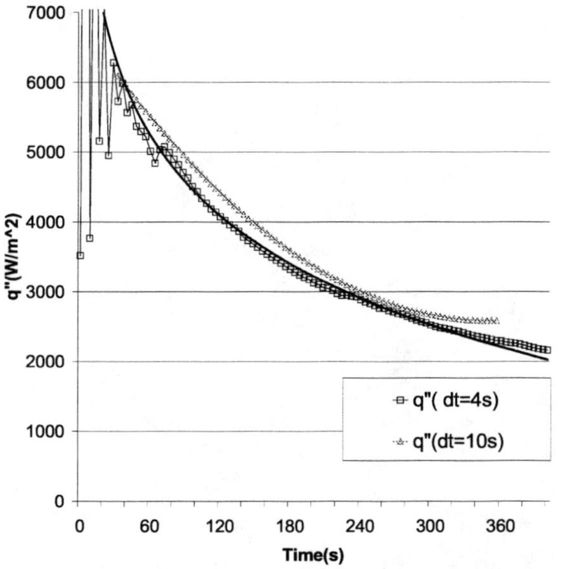

Fig. 5 Heat fluxes for both cases of silicone rubber cooled with dry ice and acetone.

are quite close, beginning at q" values of 7,000 W/m^2 and then decreasing to 2,000 – 2,500 W/m^2 near the end of the experiments. In the simulation for these two cases, the value of $\Delta x = 0.3125$ mm is the same, but the values of Δt are different. It is concluded that both time steps give good results of the transient heat transfer behavior.

The second silicon rubber experiment was performed using the dry ice and PFC mixture. The results are shown in Fig. 6, in which the heat flux curve is similar to that in the dry ice and acetone experiments. In Fig. 6, a peak heat flux value (7,000 W/m^2) similar to Fig. 5 was reached shortly after the initial pouring of the cold mixture onto the sample. However, the heat flux values drop, at a faster rate than those in Fig. 5, to a value of 1,600 W/m^2 at the end of the PFC experiment. The difference may be caused by different thermophysical properties of the two fluids (acetone and PFC) used in the experiments. Further experiments are needed to understand this discrepancy in convective heat transfer. The heat transfer coefficient with PFC injection is expected to be in the range of 1,000 W/m^{2}K to 4,000 W/m^{2}K. Because the PFC fluids are very expensive, only well planned experiments will be performed.

Transient Temperature Distribution in the Biological Model

The 1-D model was then used to predict the temperature history of each of the eight hysterectomized uteri ($\alpha = 1.8 \times 10^{-7} m^2/s$) with each uterus thickness as the input and the PFC fluid temperature as the boundary condition at x = 0 for the numerical model. The actual cold surface temperatures were 1-2 °C above the fluid temperatures due to finite convective heat

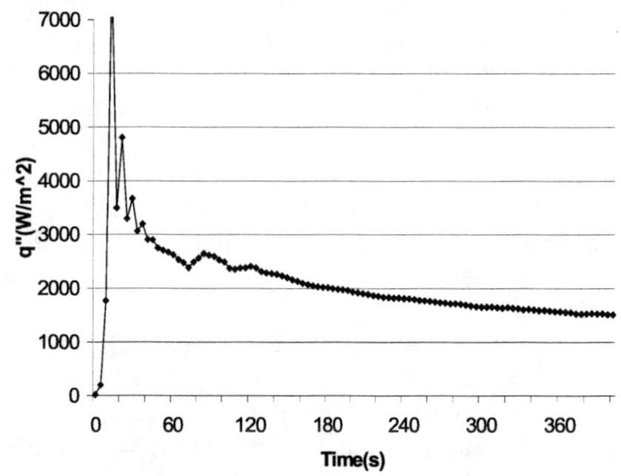

Fig. 6 Heat flux for the case of silicone rubber cooled with dry ice and PFC.

transfer, which is being investigated in our lab. The other boundary condition outside the myometrium at x = L was kept constant at 37°C in a saline bath (Fig. 2). After the experiment was completed and the temperatures recorded, each uterus was dissected longitudinally and the exact depth of the probes were measured and recorded as shown in Table 2.

Because of difficulty in resolving the depths of the thermocouples and movement during PFC infusion, the measured depth of the probe is accurate at ±1 mm of the actual depth. For all experiments, the Crank-Nicolson algorithm gave a good approximation of the measured temperature inside each of the media. There are a few reasons for the slight difference between the numerical and measured temperatures. The first reason is that the Crank Nicholson method uses many simplifications to the actual problem. For instance, the problem

Table 2 Thermocouple Location and Tissue Thickness (mm)

Patient Uterus	External Surface	Left Lateral	Right Lateral	Anterior Depth	Thick ness
1	18	10	7	8	20
2	24	12	12	4	24
3	28	21	9	11	30
4	26	8	13	13	28
5	28	10	Fell out	15	30
6	24	10	7	Fell out	25
7	17	14	15	12	18
8	22	12	14	10	24
Mean	23.4	12.1	11	10.4	24.8
Std. Dev.	4.17	4.01	3.31	3.35	4.38

is simplified to a one dimension problem. Also the thermal properties of the material are considered constant. In reality, they can vary with temperature. The other reason for the very slight difference between the Crank Nicholson prediction and the measured temperatures can be due to instrumental imperfection. As mentioned earlier, the crude nature of our experimental apparatus and the uncertainty about the exact position of the probes contribute to the slight difference between experimental data and theoretical prediction. Another reason is the phase change due to ice formation from the moisture content in the uterus.

With the proper insulation system, we could get an intracavitary PFC temperature of –80°C. A sensitivity analysis and prediction of the temperature distribution is shown in Fig. 7. The thickness of Uterus 4 was used as the input. The bottom curve is the estimated PFC temperature for six (6) consecutive injections for a total of 400 seconds. The calculated internal temperatures follow the oscillating behavior of the fluid temperature. The oscillating behavior was less pronounced with increasing depth due to diffusion. Further, the temperature gradients were larger near the cold surface (9°C/mm) and smaller at x = 7 mm (6°C/mm) at t = 300-400 s.. This simulation demonstrates that cryoablation can be successfully performed on the uterus at depth of 5 mm by keeping the temperature below -30°C for about 200 s.

For clarity, the temperature history of only one representative uterus is shown in Fig. 8. The thickness of the uterus 6 was 25 mm. The right lateral probe was 7 mm deep. The bottom three curves are the PFC inlet, outlet and average temperatures as a result of 6 injections. The average PFC time-dependent temperature was used in the FDM simulation. We noticed large variation in the fluid temperature due to

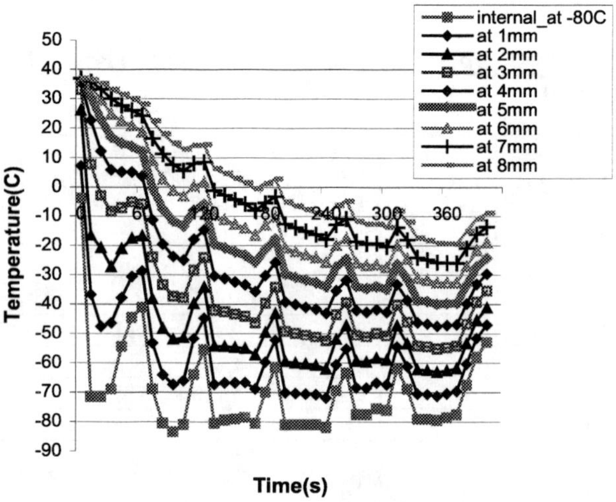

Fig. 7 Sensitivity analysis and temperature prediction for patient uterus 4

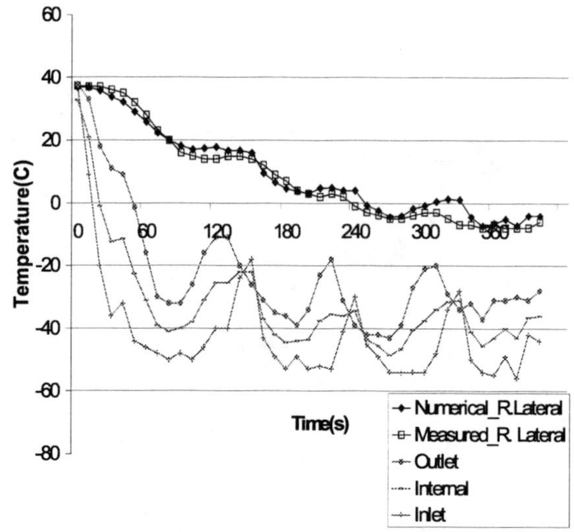

Fig. 8 Computed and measured temperature history for patient uterus 6 (thickness = 25 mm, x = 7 mm).

inadequate insulation of the injector apparatus. Nevertheless, the computed and the measured temperatures are in good agreement with each other.

Because of the uncertainty about the exact position of the temperature probes in the uterus, some experimental data agreed strongly with the numerical prediction and some did not. To make good use of data sets, we turn to statistics. Instead of using the thickness for each uterus, we decided to use the average thickness, the average location of the Right lateral probe, the average location of the Left lateral probe and the average location of the anterior probe. We also calculated the standard error for all measurements and then compared the statistical data to the numerical temperature prediction. We then compared the measured temperatures with the predicted temperatures given by the Crank Nicholson method using $\alpha=1.8.10^{-7}m^2/s$.

As shown in Fig. 9, the numerical temperature prediction is in good agreement with the mean temperature for the left lateral probe at an average depth of 12 mm and an average thickness of 23 mm. The bottom curve is the measured fluid mean temperature. The oscillations due to the 6 injections are smoothed out to some degree by the average of all patients. The numerically predicted temperatures (rectangular symbols) always stay within the range of ± one standard error of the measure temperatures, as shown by the diamond and triangular symbols.

Fig. 10 compares the predicted temperatures with the measured average temperatures for the right lateral mean probe location at an average depth of 11 mm and an average thickness of 23 mm. Fig. 11 shows the predicted temperatures with the

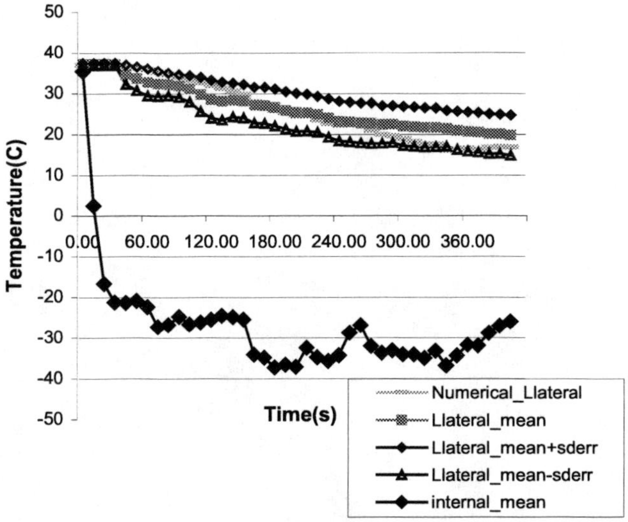

Fig. 9 Left Lateral mean probe at 12 mm and average thickness of 23 mm

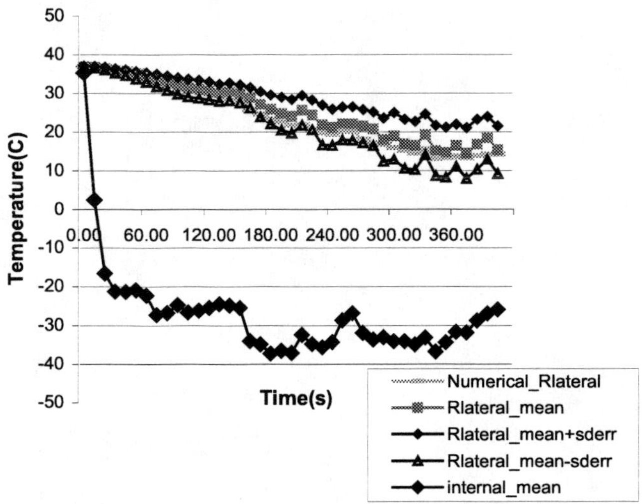

Fig. 10 Right Lateral mean probe at 11 mm and average thickness of 23 mm.

measured average temperatures for the anterior mean probe location at an average depth of 10 mm and an average thickness of 23mm. In both cases, the theoretical temperature prediction is very close to the mean temperature. In addition, the predicted temperatures always stay within a range of ± one standard error of the measure temperature in all three probe locations (Figs. 9-11). We can therefore conclude that the numerical model is a good prediction of the temperature history inside the uterus.

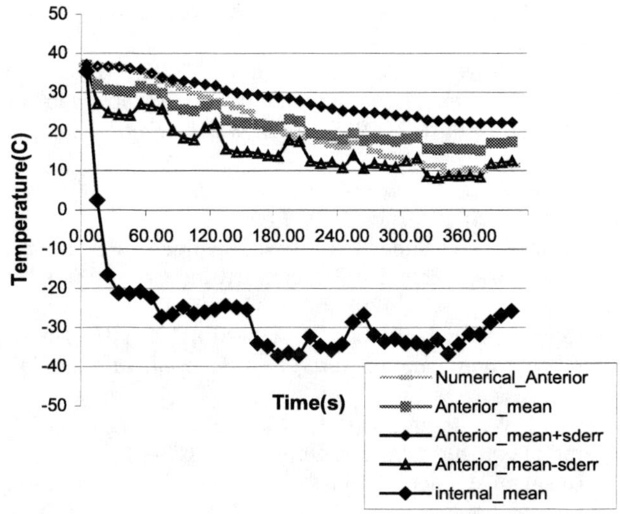

Fig. 11 Anterior mean probe at 10 mm and average thickness of 23 mm.

CONCLUSION

A one-dimensional finite difference method based on the Crank-Nicolson scheme has been developed to predict the temperature distribution inside human uteri. The model is validated using a silicon rubber sheet cooled by a dry ice-acetone mixture and then a dry ice-PFC mixture, in which the heat transfer coefficients are in the range of 800 – 4,500 W/m^2K. Good agreements are obtained by comparing the predicted temperatures with the measured temperatures at the right lateral, left lateral, and anterior locations of 10 patients whose dysfunctional uteri were removed for the *in vitro* experiments.

ACKNOWLEDGMENTS

The work was supported in part by funding from CryoFluor Therapeutics, LLC. Tucson, AZ.

REFERENCES

Best and Taylor, Physiological Basis of Medical Practice, 1991, 12th ed.. Baltimore, Williams &Wilkins.

Chato, J. C., 1985, "Measurement of thermal properties of biological materials," *Heat transfer in medicine and biology: analysis and applications*, Vol. 1, (ed. A. Shit.zer and R. C. Eberhart), pp. 167—92. Plenum Press, New York.

Cooper, J. M. and Erikson, M. L., 2000, "Global Endometrial Ablation Technologies," *Obstet and Gynecol Clinics*, pp. 385-396, 27(2).

DeCherney A. Diamond MP, Lavy G, et al., 1987, "Endometrial ablation for intractable uterine bleeding:

hysteroscopic resection," *Obstet Gynecol*:70:668.

Hartnett, J. P., Irvine, T. F., and Cho, Y. I., 1992, *Advances in Heat Transfer: Bioengineering Heat Transfer*, Vol. 22, Academic Press.

Hoffman, N. E. and Bichof, J. C., 2001, "Cryosurgery of Normal and Tumor Tissue in the Dorsal Skin Flap Chamber: Part I-Thermal Response," *ASME J. Biomech. Eng.*, Vol. 123, pp. 301-309.

Ozisik, M. N., 1994, *Finite Difference Methods in Conduction Heat Transfer*, CRC Press.

Rabin, Y. and Shitzer, A., 1998, "Numerical Solution of the Multidimensional Freezing Problem During Cryosurgery," *ASME J. Biomech. Eng.*, Vol. 120, pp. 33-37.

Shitzer, A. and Eberhart, K. C. (ed.), 1985, *Heat transfer in medicine and biology: analysis and applications*, Vol. 1. Plenum Press, New York.

Song, W. L. and Jiji, L. M., 1988, "Peripheral tissue freezing in cryosurgery," *Cryobiology*, 2, 153-63.

Tatsutani, K. and Rubinsky, B., 1998, "A Method to Study Intracellular Ice Nucleation," *ASME J. Biomech. Eng.*, Vol. 112, pp. 27-31.

Tunnel, J. W., Torres, J. H., and Anvari, B., 2002, "Methodology for Estimation of Time-Dependent Surface Heat Flux due to Cryogen Spray Cooling," *Annals of Biomedical Engineering*, Vol. 30, pp. 19-33.

Weinbaum, S., and Jiji, L. M., 1985, "A New Simplified Bioheat Equation for the Effect of Blood Flow on Local Average Tissue Temperature," *ASME J. Biomech. Eng.*, Vol. 107, pp. 131-139.

Wolfson, M., Greenspan J., and Shaffer T., 1998, "Liquid-Assisted Ventilation: An Alternative Respiratory Modality," *Ped. Pulmonology,* 26: 42-63.

AUTHOR INDEX

HTD-Vol. 373/BED-Vol. 53
Advances in Heat and Mass Transfer in Biotechnology — 2002

Book Number: I00603